Name ______________________ Class ______________ Date __

Skills Worksheet

Directed Reading

AF769477

Section: Characteristics of Fungi

Complete each statement by underlining the correct term or phrase in the brackets.

1. Fungi are [heterotrophic / autotrophic].

2. Fungal cells have cell walls made of [chitin / cellulose].

Read each question, and write your answer in the space provided.

3. What is distinctive about the body structure of fungi?

____

____

____

4. What is nuclear mitosis?

____

____

____

5. What are hyphae?

____

____

6. What is a mycelium?

____

____

7. How do fungi absorb nutrients?

____

____

____

Complete each statement by underlining the correct term or phrase in the brackets.

8. Fungi reproduce by releasing [spores / gametes] in reproductive structures at the tips of hyphae.

9. Fungal spores are [diploid / haploid].

10. In fungi, most spores are formed by mitosis during [asexual reproduction / sexual reproduction].

11. In sexual reproduction, hyphae belonging to two different [sexes / mating types] fuse together.

Name ____________________ Class ______________ Date ______________

Skills Worksheet

Directed Reading

Section: Fungal Diversity

Read each question, and write your answer in the space provided.

1. What are deuteromycetes?

____

____

____

2. What familiar fungi are considered deuteromycetes?

____

____

____

3. Truffles are a member of which phylum? What are the distinctive characteristics of this phylum?

____

____

____

4. What is the difference between the hyphae of Ascomycota and Zygomycota?

____

____

____

5. Mushrooms are a member of which phylum? What are the distinctive characteristics of this phylum?

____

____

____

In the space provided, write the letter of the description that best matches the term or phrase.

______ **6.** zygosporangia

______ **7.** stolons

______ **8.** rhizoids

a. mycelia that grow along the surface of a food source

b. mycelia that anchor the fungus in a food source

c. thick-walled sexual structures

Complete each statement by writing the correct term or phrase in the space provided.

9. The ________________________ is a saclike structure in which haploid spores are formed.

10. The common name given to unicellular ascomycetes is ________________________ .

11. In a process called ________________________ , a small cell forms from a large cell and pinches itself off from the large cell.

Complete each statement by underlining the correct term or phrase in the brackets.

12. Basidiomycetes usually reproduce [asexually / sexually].

13. The basidium is a club-shaped [reproductive / nutrient-collecting] structure.

14. Some rusts and smuts reproduce [sexually / asexually].

15. *Amanita muscaria* is [poisonous / safe to eat].

Name ______________________ Class ______________ Date ____________

Skills Worksheet

Directed Reading

Section: Fungal Partnerships

Read each question, and write your answer in the space provided.

1. What is a mycorrhiza?

____

____

2. What benefit do fungi derive from mycorrhizae?

____

____

3. What benefit do plants derive from mycorrhizae?

____

____

____

Complete each statement by underlining the correct term or phrase in the brackets.

4. For most plants, fungi [penetrate / wrap around] the roots in mycorrhizae.

5. When the fungus wraps around the roots in mycorrhizae, the fungus is usually a [basidiomycete / zygomycete].

Read each question, and write your answer in the space provided.

6. What is a lichen?

____

____

____

____

____

7. What does the fungus provide the photosynthetic partner in a lichen?

____

____

____

____

____

8. Why are lichens able to thrive in harsh habitats?

____

____

____

____

____

Name ______________________ Class ______________ Date ____________

Skills Worksheet

Active Reading

Section: Characteristics of Fungi

Read the passage below. Then answer the questions that follow.

Traditionally, biologists grouped fungi with plants because fungi are immobile, have a cell wall, and appear "rooted" in the soil. However, the unique features of fungi indicate that they should be classified as a separate kingdom.

Fungi are heterotrophic: Plants appear green because they contain chlorophyll, the pigment that aids in photosynthesis. Fungi do not appear green because they do not contain chlorophyll. For example, the stalk and cap of the mushroom are not green like the stem and leaves of a plant. Fungi are heterotrophs that obtain energy by breaking down organic molecules that they absorb from their environment.

Fungi have filamentous bodies: Plants are made of many cell and tissue types, but fungi are made of long, slender filaments. These filaments weave tightly together to form the fungus body and reproductive structures.

Fungal cells contain chitin: The cells of mushrooms, like the cells of all fungi, have walls made of **chitin,** the tough polysaccharide found in the hard outer covering of insects. Plant cells have walls made of cellulose, a different polysaccharide.

Fungi exhibit nuclear mitosis: Mitosis in fungi is different from mitosis in plants and most other eukaryotes. In most eukaryotes, the nuclear envelope disintegrates in prophase and re-forms in telophase. In dividing mushroom cells, however, the nuclear envelope remains intact from prophase to anaphase. As a result, spindle fibers form within the nucleus. The spindle fibers then drag chromosomes to opposite poles of the nucleus rather than to opposite poles of the cell. Mitosis is complete when the nuclear envelope pinches in two.

SKILL: READING EFFECTIVELY

Read each question, and write your answer in the space provided.

1. In the past, why did biologists classify fungi with plants?

____

____

2. Why does a mushroom lack the green coloring found in plants?

____

____

3. How do fungi obtain the energy needed to carry out life processes?

____

____

4. How does mitosis in fungi differ from that in plants and most other eukaryotes?

____

____

SKILL: RECOGNIZING SIMILARITIES AND DIFFERENCES

In the spaces provided, write *P* if the statement describes plants, write *F* if the statement describes fungi, or write *B* if the statement describes both plants and fungi.

______ **5.** made of long, slender filaments

______ **6.** lack chlorophyll

______ **7.** mitosis occurs in the organism

______ **8.** autotrophs

______ **9.** cells walls contain chitin

______ **10.** need energy to carry out life processes

______ **11.** cell walls contain cellulose

______ **12.** made of many cell and tissue types

______ **13.** immobile

______ **14.** spindle fibers form within the nucleus

In the space provided, write the letter of the term or phrase that best answers the question.

______ **15.** Which of the following is a trait shared by fungi and plants?
- **a.** exhibit nuclear mitosis
- **b.** cell walls contain a polysaccharide
- **c.** heterotrophic
- **d.** contain a variety of tissue types

Name ______________________________ Class ______________ Date ____________

Skills Worksheet

Active Reading

Section: Fungal Diversity

Read the passage below. Then answer the questions that follow.

If you place an uncovered loaf of bread near a windowsill, after a few days a cottony mold will cover its surface. This common black bread mold is *Rhizopus stolonifer*, a member of the phylum Zygomycota. Members of the phylum Zygomycota are named for the thick-walled sexual structures called **zygosporangia** that characterize the phylum's members. Species of *Rhizopus* and other zygomycetes usually live in the soil and feed on decaying plant and animal matter. The mycelia that grow along the surface of bread are called **stolons.** The hyphae that anchor the fungus in the bread are called **rhizoids.**

Asexual reproduction in zygomycetes is much more common than sexual reproduction. During asexual reproduction, haploid spores are produced in the tips of specialized hyphae. When these spores mature, they are shed and are carried by the wind to new locations, where they germinate and start new mycelia.

SKILL: SEQUENCING INFORMATION

Sequence the statements to show what occurs during asexual reproduction in zygomycetes, beginning with the formation of haploid spores. Write *A* in front of the event that occurs first, write *B* to show what happens next, and so on.

______ **1.** Wind carries spores to new locations.

______ **2.** New mycelia are produced.

______ **3.** Specialized hyphae produce haploid spores.

______ **4.** Spores germinate.

______ **5.** Mature spores are shed.

An analogy is a comparison. In the space provided, write the letter of the term or phrase that best completes the analogy.

______ **6.** Rhizoids are to hyphae as stolons are to

a. cell walls.
b. mycelia.
c. spores.
d. mold.

Name ______________________________ Class ______________ Date ____________

Skills Worksheet

Active Reading

Section: Fungal Partnerships

Read the passage below. Then answer the questions that follow.

A **lichen** is a symbiosis between a fungus and a photosynthetic partner such as a green alga, a cyanobacterium, or both. The photosynthetic partner provides carbohydrates. It is protected from the environment by the fungal partner, which helps it absorb mineral nutrients. In most lichens, the fungus is an ascomycete. When you look at a lichen, you are seeing the fungus. The photosynthetic partner is hidden between layers of hyphae. Sunlight penetrates the layers of hyphae and enables photosynthesis. The tough construction of the fungus combined with the photosynthetic abilities of the alga have enabled lichens to colonize harsh habitats. Lichens have been found in arid desert regions and in the Arctic. Lichens grow on bare soil, on tree trunks, and on sun-baked rocks. During succession, lichens are often the first colonists. They break down rocks and prepare the environment for other organisms. Lichens are a key component of primary succession because they are able to carry out nitrogen fixation.

SKILL: RECOGNIZING CAUSE AND EFFECT

Two independent events can be linked in a cause-and-effect relationship. The first event to occur, or the cause, triggers a second event, or the effect, to happen. In the spaces below, identify and write the missing part of each cause-and-effect relationship.

1. Cause: Sunlight penetrates layers of hyphae.

Effect: ____

2. Cause: ____

Effect: Lichens are able to survive harsh environments.

In the space provided, write the letter of the phrase that best completes the statement.

______ **3.** In a lichen, the fungal partner

a. protects the photosynthetic partner.
b. supplies the photosynthetic partner with mineral nutrients.
c. provides the photosynthetic partner with carbohydrates.
d. Both (a) and (b)

Name ______________________________ Class ______________ Date ____________

Skills Worksheet

Vocabulary Review

In the space provided, write the letter of the description that best matches the term or phrase.

______ **1.** chitin

______ **2.** hyphae

______ **3.** mycelium

______ **4.** zygosporangium

______ **5.** stolon

______ **6.** rhizoid

______ **7.** ascus

______ **8.** yeasts

______ **9.** budding

______ **10.** basidium

______ **11.** mycorrhizae

______ **12.** lichen

a. a type of mutualistic relationship formed between fungi and the roots of vascular plants

b. a thick-walled sexual structure

c. the tough polysaccharide found in the hard outer covering of insects and fungal cell walls

d. a symbiosis between a fungus and a photosynthetic partner

e. the hyphae that anchor a fungus to its source of food

f. slender filaments that compose the body of a fungus

g. tangled mass formed by hyphae

h. the mycelia that grow along the surface of a fungus's food

i. a saclike structure in which haploid spores are formed

j. the common name given to unicellular ascomycetes

k. a club-shaped sexual reproductive structure

l. asexual reproduction in which a small cell forms from a larger cell by pinching itself off from the larger cell

Name ______________________ Class ______________ Date ______________

Skills Worksheet

Science Skills

Classifying

Many years ago biologists grouped fungi and plants together because they appeared to share many similarities. However, fungi and plants are very different. Show the differences between a mushroom (a fungus) and an oak tree. Using the terms and phrases below, complete the table that follows. Some terms or phrases may be used more than once.

- eukaryotic
- photosynthetic
- heterotrophic
- has cell walls
- appears "rooted" in the soil
- has roots
- has rhizoids
- absorbs organic nutrients from the environment
- contains chlorophyll
- contains chitin
- has hyphae
- contains cellulose
- body of long, slender filaments
- many cell types and tissue types
- immobile

Oak tree	Mushroom	Both oak tree and mushroom

Name ______________________ Class ______________ Date ____________

Skills Worksheet

Concept Mapping

Using the terms and phrases provided below, complete the concept map showing the characteristics of fungi.

ascomycetes	morels	sexually
asexually	mushrooms	truffles
deuteromycetes	puffballs	zygomycetes

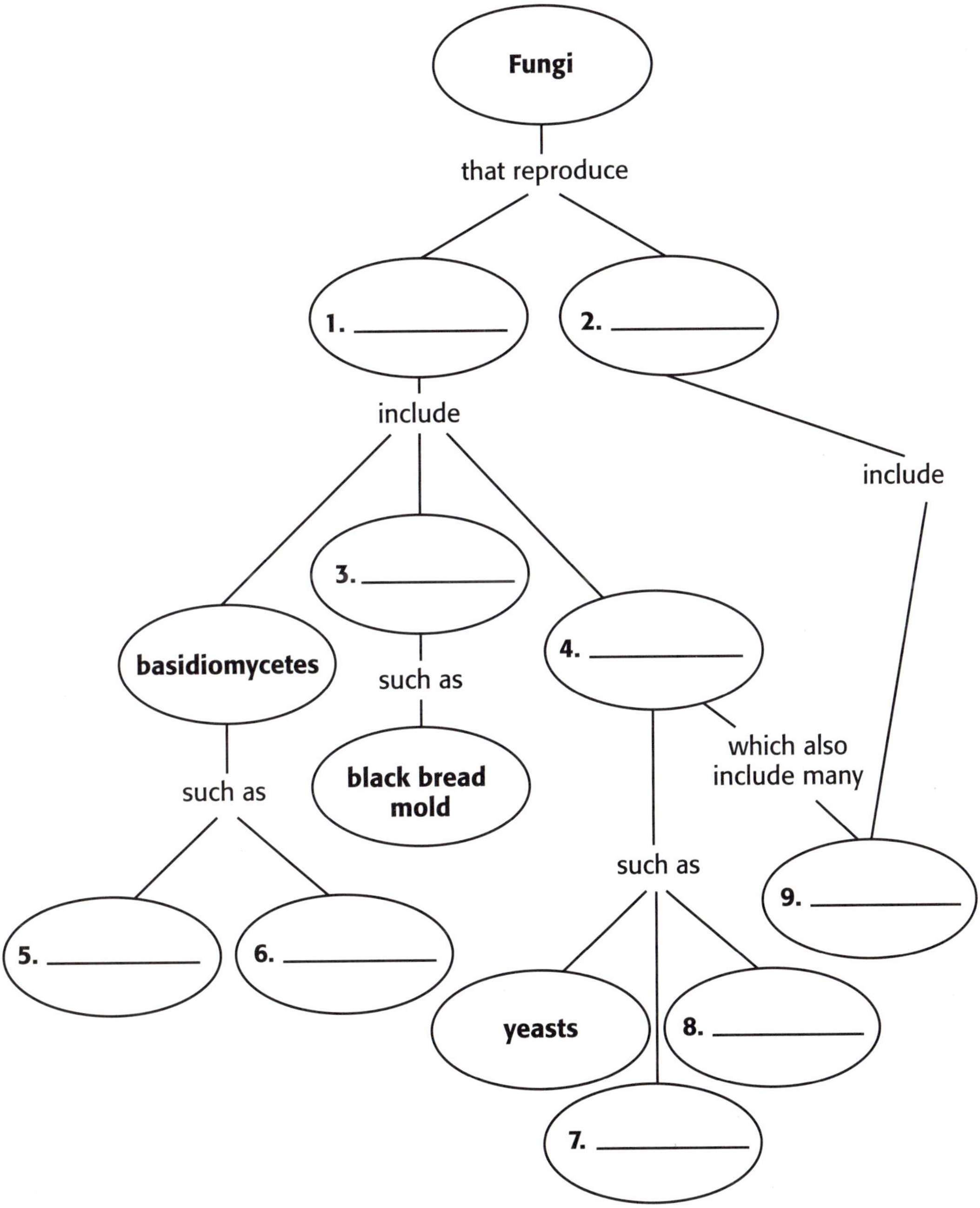

Name ______________________ Class ______________ Date ______________

Skills Worksheet

Critical Thinking

Work-Alikes

In the space provided, write the letter of the term or phrase that best describes how each numbered item functions.

______ **1.** zygosporangia

______ **2.** *Amanita* species

______ **3.** alga and fungus in lichens

______ **4.** no lichens in industrial areas

a. pollution indicator

b. food poisoning

c. thick walls

d. partners

Cause and Effect

In the space provided, write the letter of the term or phrase that best matches each cause or effect given below.

Cause	Effect
5. many fungi decompose leaves, branches, dead animals, and waste	______________
6. large, expansive mycelium of fungus	______________
7. algal photosynthesis and tough fungal construction	______________
8. ______________	possible role in invasion of land plants
9. non-penetrating mycorrhizae	______________

a. pines, oaks, beeches, and willows benefit

b. lichens' ability to live in harshest habitats

c. high surface area-to-volume relationship

d. recycling of nutrients

e. mycorrhizae in fossils of earliest land plants rootlike structures

Trade-offs

In the space provided, write the letter of the bad news item that best matches each numbered good news item below.

Good News

______ **10.** Yeasts have played an important role in brewing and baking.

______ **11.** The deuteromycetes are a fourth phylum with no sexual stages.

______ **12.** Deuteromycetes fungi are sources of the antibiotic penicillin.

Bad News

a. *C. albicans* is a species that causes thrush in humans.

b. Others in this group cause ringworm and athlete's foot.

c. Most members of this group have been reclassified into the Ascomycetes.

Linkages

In the spaces provided, write the letters of the two terms or phrases that are linked together by the term or phrase in the middle. The choices can be placed in any order. Some choices may be used more than once.

13. ______ spindle fibers drag chromosomes to opposite sides ______

14. ______ organic matter broken down ______

15. ______ spores are shed and carried by wind ______

16. ______ hyphae from two mating types fuse ______

a. fungi secrete digestive enzymes

b. fungi absorb decomposed molecules

c. zygomycetes and ascomycetes spores are produced by asexual reproduction

d. fusion makes reproductive structures and spores

e. mitosis begins in the fungal nucleus

f. they may germinate and start new mycelia

g. sexual reproduction in fungus begins

h. nuclear envelope pinches in two

Analogies

An analogy is a relationship between two pairs of terms or phrases written as a : b :: c : d. The symbol : is read as "is to," and the symbol :: is read as "as." In the space provided, write the letter of the pair of terms or phrases that best completes the analogy shown.

______ **17.** fungal enzyme : organic matter ::
- **a.** soap : dirt
- **b.** blanket : wristwatch
- **c.** backpack : book
- **d.** paper : pencil

______ **18.** asexual reproduction : zygomycetes ::
- **a.** thick-walled reproductive structures : ascomycetes
- **b.** sac reproductive structure : basidiomycetes
- **c.** sexual reproduction : ascomycetes
- **d.** sexual reproduction : basidiomycetes

______ **19.** basidiomycetes : club ::
- **a.** ascomycetes : sac
- **b.** zygomycetes : bowling pin
- **c.** deuteromycetes : foot
- **d.** ascomycetes : golf ball on a tee

______ **20.** basidiomycetes : basidia
- **a.** zygomycetes : asci
- **b.** ascomycetes : zygospores
- **c.** deuteromycetes : basidia
- **d.** ascomycetes : asci

Name ______________________ Class ______________ Date ____________

Skills Worksheet

Test Prep Pretest

In the space provided, write the letter of the term or phrase that best completes each statement or best answers each question.

______ **1.** Which of the following is NOT a characteristic of fungi?
a. filamentous bodies
b. cell walls made of chitin
c. chlorophyll
d. nuclear mitosis

______ **2.** A mycelium helps a fungus absorb nutrients from its environment because it provides
a. minerals.
b. a high surface area-to-volume ratio.
c. digestive enzymes.
d. a low surface area-to-volume ratio.

Questions 3 and 4 refer to the figure at right.

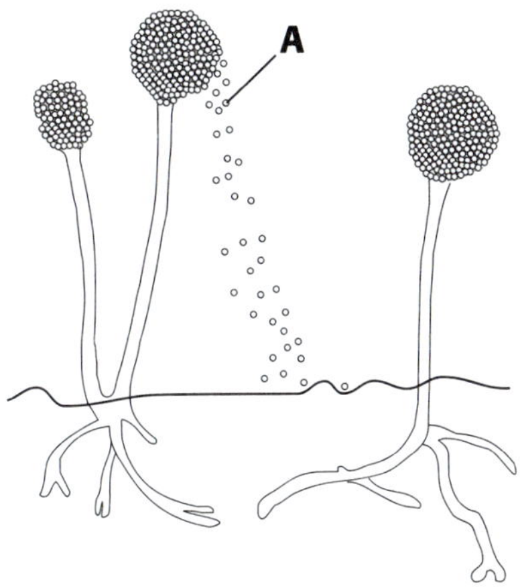

______ **3.** The fungus shown is a(n)
a. asomycete.
b. basidiomycete.
c. deuteromycete.
d. zygomycete.

______ **4.** The structure labeled *A* in the figure is called a
a. rhizoid.
b. spore.
c. stolon.
d. hypha.

______ **5.** The classification of organisms in the three phyla of the kingdom Fungi is based on
a. food.
b. digestive structures.
c. cellular structure.
d. sexual reproductive structures.

______ **6.** Most fungal spores are formed by
a. the fusing of hyphae.
b. the fusing of asci.
c. mitosis.
d. None of the above

Complete each statement by underlining the correct term or phrase in the brackets.

7. The tough material found in the cell walls of all fungi is [cellulose / chitin].

8. The slender filaments that make up the bodies of most fungi are called [hyphae / mycelium].

9. In bread mold, the mycelia that grow along the surface of the bread are called [stolons / rhizoids].

10. During succession, [mycorrhizae / lichens] are often the first living organisms to appear in a new area.

11. In lichens, the algal partner provides [minerals / carbohydrates].

Complete each statement by writing the correct term or phrase in the space provided.

12. When you look at a lichen, you are looking at the ______________________ , which is usually a(n) ______________________ .

13. Fungi secrete ______________________ ______________________ that break down organic matter so they can then absorb the decomposed molecules.

14. Sexual reproduction in fungi is initiated when two ______________________ of opposite mating types fuse and form a reproductive structure.

15. The three phyla of fungi form distinctive structures during sexual reproduction. Members of the phylum Zygomycota form ______________________ ; members of the phylum Ascomycota form ______________________ ; and members of the phylum Basidiomycota form ______________________ .

16. Fungi ______________________ nonliving organic matter and thus are resource recyclers.

17. Certain fungi play important roles in the nutrition of vascular plants by forming symbiotic associations with their roots, called ______________________ .

18. The underside of a mushroom cap is lined with rows of ______________________ , which contain thousands of club-shaped structures called ______________________ .

19. Fungi that are ______________________ compete for nutrients with their hosts.

20. Fungi called ______________________ are commercially useful in baking, brewing, and wine-making.

Read each question, and write your answer in the space provided.

21. Give four reasons fungi are no longer classified in the same kingdom as plants.

22. Distinguish between mitosis in fungi and mitosis in plants and most other eukaryotes.

23. Why can lichen survive in habitats as diverse as polar and desert regions?

24. Give at least two examples of a beneficial and a harmful ascomycete.

25. Describe how *Amanita muscaria* obtains nutrients.

Name ______________________ Class ______________ Date ____________

Assessment

Quiz

Section: Characteristics of Fungi

In the space provided, write the letter of the description that best matches the term or phrase. Some choices may be used more than once.

______ **1.** grows on bread

______ **2.** causes athlete's foot

______ **3.** source of certain antibiotics

______ **4.** breaks down fallen leaves and tree stumps

______ **5.** causes ringworm

______ **6.** baker's yeast

______ **7.** produces flavor and aroma of certain cheeses

a. parasite that absorbs nutrients from living host

b. resource recycler

c. commercially valuable

In the space provided, write the letter of the term or phrase that best completes each statement or best answers each question.

______ **8.** A typical fungus
- **a.** contains chlorophyll.
- **b.** is immobile.
- **c.** has cell walls made of chitin.
- **d.** Both (b) and (c)

______ **9.** Fungi obtain nutrients by
- **a.** taking in food particles and digesting them internally.
- **b.** capturing microorganisms that live in the soil.
- **c.** breaking down organic material in their environment and digesting it.
- **d.** using sunlight to make carbohydrates.

______ **10.** Fungi reproduce by releasing which of the following?
- **a.** lightweight spores
- **b.** thick-walled resistant zygotes
- **c.** very small seeds
- **d.** fast-swimming sperm

Name ______________________________ Class ______________ Date ____________

Assessment

Quiz

Section: Fungal Diversity

In the space provided, write the letter of the description that best matches the term or phrase.

______ 1. *Aspergillus*

______ 2. *Penicillium*

______ 3. *Rhizopus*

______ 4. *Saccharomyces*

______ 5. *Amanita*

______ 6. *Candida*

a. baker's yeast

b. deadly mushroom

c. causes thrush

d. source of an antibiotic

e. used to ferment soy sauce and to produce citric acid

f. bread mold

In the space provided, write the letter of the term or phrase that best completes each statement or best answers each question.

______ 7. Sexual reproduction occurs by means of a fruiting body called a mushroom in
a. ascomycetes.
b. yeasts.
c. zygomycetes
d. basidiomycetes.

______ 8. Rusts are fungi that
a. attack crop plants.
b. oxidize unprotected iron.
c. enable cheese to ripen faster.
d. are used in commercial wine-making.

______ 9. Asexual reproduction by means of shedding haploid spores is rare among the
a. zygomycetes.
b. ascomycetes.
c. basidiomycetes.
d. deuteromycetes.

______10. Which group is composed of fungi in which no sexual stage has been seen?
a. zygomycetes
b. ascomycetes
c. basidiomycetes
d. deuteromycetes

Name ______________________________ Class ______________ Date ____________

Assessment

Quiz

Section: Fungal Partnerships

In the space provided, write the letter of the description that best matches the term or phrase. Some choices may be used more than once.

______ **1.** include a photosynthetic partner

______ **2.** a symbiotic relationship between two living things

______ **3.** found underground

______ **4.** one partner is a fungus

______ **5.** one partner is a plant (often a tree)

______ **6.** neither partner benefits

______ **7.** includes an alga

a. mycorrhizae
b. lichens
c. both mycorrhizae and lichens
d. neither mycorrhizae nor lichens

In the space provided, write the letter of the term or phrase that best completes each statement or best answers each question.

______ **8.** Plants can grow successfully in infertile soils if the plants have which of the following?
a. truffles
b. mushrooms
c. lichens
d. mycorrhizae

______ **9.** In mycorrihizae, the hyphae penetrate the outer cells of a plant's
a. leaves.
b. roots.
c. stems.
d. fruits.

______ **10.** The algae in lichens provide the ability to use which of the following to make carbohydrates?
a. water
b. air
c. chitin
d. sunlight

Name ______________________ Class ______________ Date ____________

Assessment

Chapter Test

Fungi

In the space provided, write the letter of the fungal group that best matches the term or phrase. Some choices may be used more than once.

______ **1.** bread molds

______ **2.** yeasts

______ **3.** mushrooms

______ **4.** morels and truffles

______ **5.** rusts

______ **6.** athlete's foot

______ **7.** puffballs

______ **8.** ringworm

______ **9.** thrush

______**10.** death angel

a. Zygomycota
b. Ascomycota
c. Basidiomycota
d. Deuteromycota

In the space provided, write the letter of the term or phrase that best completes each statement or best answers each question.

______**11.** All fungi except yeasts have bodies composed of slender filaments called
a. stems.
b. hyphae.
c. roots.
d. stalks.

______**12.** Commercial uses for fungi include which of the following?
a. antibiotics
b. producing flavors for cheese
c. wine-making
d. All of the above

______**13.** Unlike animals, fungi
a. are heterotrophic.
b. are nonmobile.
c. have cell membranes.
d. are photosynthetic.

______**14.** The recognizable part of *Amanita muscaria* is the
a. mushroom.
b. bread mold.
c. septa.
d. ascus.

______**15.** The type of reproduction used by yeasts, in which a small cell forms from a large cell and pinches itself off from the large cell, is called
a. conjugation.
b. fusion.
c. budding.
d. None of the above

______**16.** Sexual reproduction occurs by means of a fruiting body called a mushroom in
a. ascomycetes.
b. yeasts.
c. basidiomycetes.
d. zygomycetes.

______**17.** The fungi in mycorrhizae provide minerals and other nutrients that they absorb from the
a. air.
b. soil.
c. infected plant.
d. alga.

______**18.** Fungi that decompose nonliving organic matter are considered to be
a. infectious.
b. parasitic.
c. pathogenic.
d. recyclers.

______**19.** A symbiosis between a fungus and a photosynthetic partner is a
a. lichen.
b. mycorrhiza.
c. moss.
d. Both (a) and (b)

______**20.** Which of the following fungi causes an infectious human disease?
a. *Saccharomyces*
b. *Candida*
c. *Penicillium*
d. *Rhizopus*

Name ______________________ Class ______________ Date ____________

Assessment

Chapter Test

Fungi

In the space provided, write the letter of the term or phrase that best completes each statement or best answers each question.

______ **1.** Fungi play an important role in the biosphere because they
a. break down organic molecules.
b. help recycle nutrients.
c. are decomposers.
d. All of the above

______ **2.** The individual filaments that make up the body of a fungus are called
a. vascular tissue.
b. hyphae.
c. rhizoids.
d. stems.

______ **3.** Fungi
a. do not contain chloroplasts.
b. have cell walls that contain chitin.
c. do not produce their own food.
d. All of the above

______ **4.** In both phylum Ascomycota and phylum Basidiomycota,
a. sexual spores form in basidia.
b. hyphae are divided by walls.
c. sexual spores form in asci.
d. the mycelia are called stolons.

______ **5.** *Amanita muscaria* has which of the following?
a. cell walls made of chitin
b. club-shaped reproductive cells called asci
c. the polysaccharide cellulose
d. All of the above

______ **6.** Fungi are used to
a. produce antibiotics.
b. ferment soy sauce.
c. flavor cheese.
d. All of the above

______ **7.** Mycorrhizae
a. aid in the transfer of minerals from the soil to a plant.
b. cause a variety of plant diseases.
c. aid in the transfer of minerals to fungi.
d. are found only on aquatic fungi.

______ **8.** Asexual reproduction is rare among the
a. zygomycetes.
b. ascomycetes.
c. basidiomycetes.
d. deuteromycetes.

Questions 9–11 refer to the figure below, which shows the life cycle of a mold.

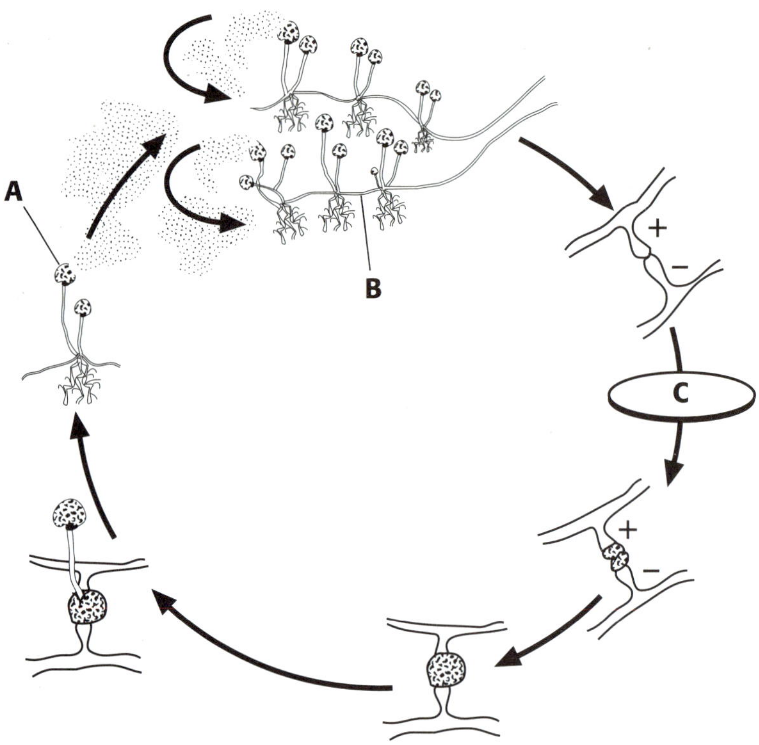

______ **9.** The structure labeled *A* produces
a. asci.
b. haploid spores.
c. rhizoids.
d. stolons.

______ **10.** The structure labeled *B* is
a. a rhizoid.
b. vascular tissue.
c. a stolon.
d. a stem.

______ **11.** The process that takes place at stage *C* is known as
a. meiosis.
b. conjugation.
c. mitosis.
d. fusion.

Complete each statement by writing the correct term or phrase in the space provided.

12. Some fungi obtain food by decomposing ______________________

______________________ in dead or living organisms.

13. A fungal ______________________ is a haploid reproductive cell that is capable of developing into a new organism.

14. A(n) ______________________ is a saclike structure in which haploid spores are formed.

15. A(n) ______________________ of a fungus, made of a tangled mass of ______________________ , is usually hidden within the substrate on which the fungus is growing.

16. Based on the structures produced during sexual reproduction fungi can be classified into three phyla of fungi and these phyla are ______________________ , ______________________ , and ______________________ .

17. Fungi that absorb nutrients in a person's body can cause life-threatening ______________________ .

18. Fungi exhibit ______________________ mitosis, in which the nuclear envelope remains intact from prophase to anaphase.

19. The unicellular fungi used in the production of baked goods and the production of alcoholic beverages are called ______________________ .

20. Fungi in which no sexual stage has been seen are classified in a group called the ______________________ .

21. Biologists use the relative health of ______________________ and their chemical compositions as indicators of air pollutants.

Read each question, and write your answer in the space provided.

22. Both mycorrhizae and lichens have symbiotic relationships. Describe how they differ from each other.

____

____

____

____

____

____

____

____

23. Why are fungi well suited for absorbing food from the environment?

____

____

____

____

24. Explain how fungi in the phylum zygomycota typically reproduces..

____

____

____

____

25. How did mycorrhizae help plants invade the land?

____

____

____

____

Name ______________________ Class ______________ Date ____________

Quick Lab

DATASHEET FOR IN-TEXT LAB

Observing Bread Mold

You can use a microscope to see the individual threads of cells that make up the body of a fungus.

MATERIALS

- prepared slide of *Rhizopus*—black bread mold
- compound microscope

Procedure

1. Examine a slide of black bread mold under low power of a microscope.
2. Move the slide to an area where you can clearly see threadlike structures.
3. Draw what you see in your lab notebook. Be sure to use at least one-third of the page.
4. Move the slide to examine an area where you can clearly see the round bulb-like structures.
5. Draw what you see in your notebook.

Analysis

1. **Label** the drawings you made, including the hyphae and mycelium.

   ____

2. **Explain** where you would find each structure on a loaf of bread.

   ____

   ____

3. **Critical Thinking**
   Recognizing Relationships Relate the structures you drew to their functions described in your textbook.

   ____

   ____

   ____

Name ______________________ Class ______________ Date ____________

Data Lab

DATASHEET FOR IN-TEXT LAB

Analyzing the Effect of Mycorrhizae

Background

Two groups of plants were planted in similar soils under similar conditions, but group A was grown in sterilized soil and group B was grown in nonsterilized soil. After 18 weeks of growth, a photograph was taken of the plants. Examine the drawing, and answer the following questions:

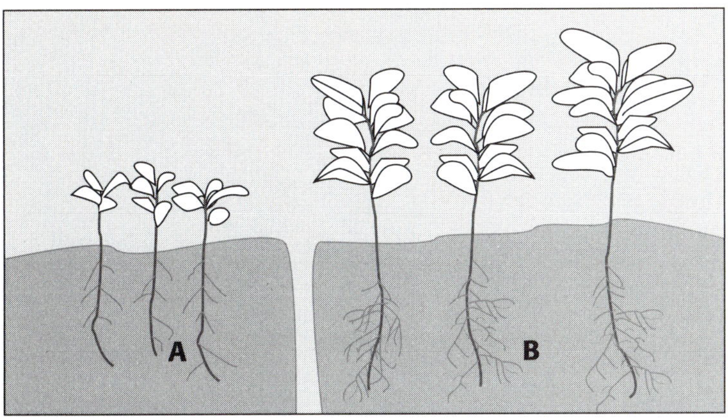

Analysis

1. **Compare** the growth of the two groups. Which grew faster?

____

2. **Explain** why one group grew better than the other group.

____

____

3. **Critical Thinking**
Inferring Relationships Suggest a possible cause of slower growth in the smaller plants.

____

____

4. **Recommend** a course of action to restore growth in the stunted plants.

____

Name ______________________ Class ______________ Date ____________

Skills Practice Lab | **DATASHEET FOR IN-TEXT LAB**

Observing Yeast and Fermentation

SKILLS

- Observing
- Measuring
- Collecting data
- Analyzing data

OBJECTIVE

- **Observe** the release of energy by yeast during fermentation.

MATERIALS

- 500 mL vacuum bottle
- 10 cm glass tubing
- 2-hole rubber stopper
- 250 mL beaker
- 75 g sucrose
- one package dry yeast
- thermometer
- 50 cm rubber tubing
- 150 mL limewater

CHEMSAFETY

CAUTION: Always wear safety goggles and a lab apron to protect your eyes and clothing.

CAUTION: Glassware is fragile. Notify the teacher of broken glass or cuts. Do not clean up broken glass or spills with broken glass unless the teacher tells you to do so.

Before You Begin

Sucrose is a disaccharide—a carbohydrate made from two monosaccharides. It is one chemical made by plants to store the sun's energy. Yeast release the energy stored in sucrose in a process called **fermentation.** In this investigation you will have a chance to observe and measure the products of fermentation.

1. Write a definition for the boldface term in the paragraph above.

____

____

2. You will be using the data table below to record your data.

Data Table		
Fermentation by Yeast		
Time Date Temp.	**Time Date Temp.**	**Time Date Temp.**
1.	8.	15.
2.	9.	16.
3.	10.	17.
4.	11.	18.
5.	12.	19.
6.	13.	20.
7.	14.	21.

Procedure

1. Set up your vacuum bottle according to the diagram.

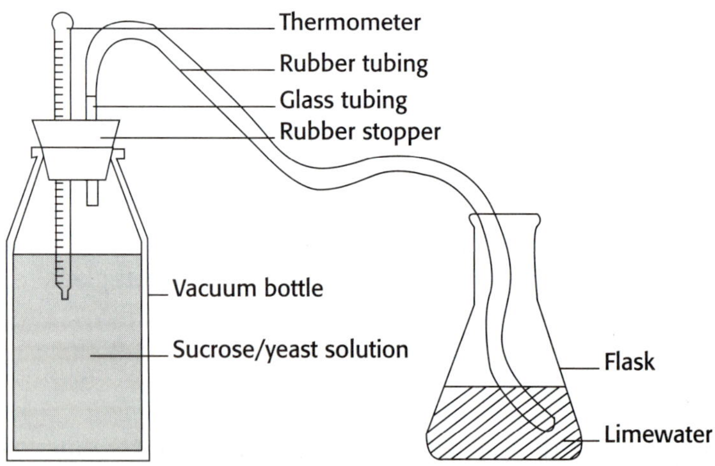

2. Mix 75 g of sucrose in 400 mL of water.

3. When the sucrose has dissolved, add one-half package of fresh yeast and stir.

4. Pour the sucrose-yeast solution into a vacuum bottle until it is approximately three-quarters full.

5. Adjust the thermometer so that it extends down into the sugar-yeast solution.

6. Record the temperature of the solution on the observation chart as soon as possible. Continue to record the temperature as often as possible during the next two days.

Analyze and Conclude

1. Summarizing Results Prepare a graph of your data, illustrating the temperature over time. Complete the graph by drawing a curve through the plotted points.

____

____

2. Analyzing Data What does the curved line plotted on the graph indicate?

____

____

3. Drawing Conclusions What can you conclude about the energy contained in sucrose?

____

____

____

____

____

4. Predicting Patterns What do you think would happen if there were only one hole in the stopper for the thermometer?

____

____

5. Further Inquiry If you know that fermentation liberates energy and gives off carbon dioxide and alcohol as waste products, how would you prove that fermentation is really taking place in the sugar-yeast solution?

____

____

____

Name ________________________________ Class ________________ Date ______________

Exploration Lab

OBSERVATION

Exploring Fungi

Fungi are classified into four phyla based on their sexual reproductive structures. There are three phyla of sexually reproducing fungi: Zygomycota, Ascomycota, and Basidiomycota. Each group is characterized by distinctive sexual reproductive structures.

FIGURE 1 THREE PHYLA OF FUNGI

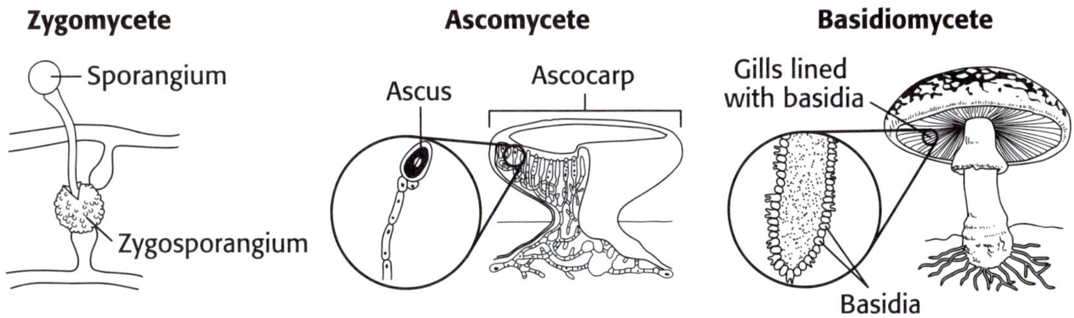

Zygomycetes form thick-walled sexual structures called *zygosporangia*. You may have seen zygomycetes growing on old bread in the form of mold. *Ascomycetes*, which include yeasts and truffles, can reproduce sexually and asexually. They produce sexual spores in the form of saclike *asci*. Basidiomycetes, which include mushrooms, form club-shaped *basidia* as sexual spores. The phylum called Deuteromycota includes more than 17,000 fungi, which appear to reproduce only asexually.

In this lab, you will examine and compare specimens from the four fungi phyla.

OBJECTIVES

Observe a specimen from each phylum of fungus.

Draw and label pictures of a representative sample from each phylym of fungus.

Identify where each type of fungus might be found and what it uses as a food source.

MATERIALS

- compound light microscope
- field guide to fungi
- lab apron
- representative slides of each phylum of fungi
- safety goggles
- transparent tape

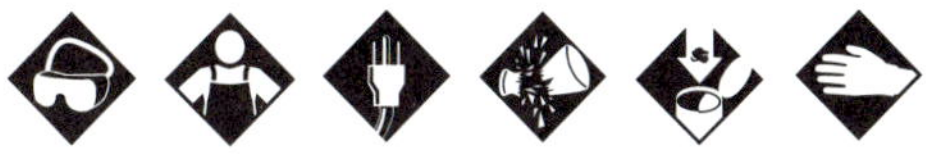

Procedure

1. Choose one of the slides provided by your teacher. Record your observations about the fungi appearance in **Table 1.**

TABLE 1 OBSERVATIONS OF FOUR PHYLA OF FUNGI

Phylum and specimen name	Appearance	Drawing of reproductive structures	Location and food sources
Zygomycota Culture: ______________			
Ascomycota Culture: ______________			
Basidiomycota Culture: ______________			
Deuteromycota Culture: ______________			

2. Locate the reproductive area of the fungus. If you cannot find the reproductive structures, look in your textbook or ask your teacher for help.

3. Observe the reproductive structures by using the low power of the microscope. Record your observations by drawing and labeling what you see. If your drawing will not fit in the space provided in **Table 1,** then use a separate sheet of paper.
4. Use your textbook, a field guide, or other resource to determine where this type of fungus might be found. What are some possible food sources for this type of fungus? Record your answers in **Table 1.**
5. Repeat steps 1–4 for each slide provided.
6. Dispose of your materials according to your teacher's instructions.
7. Clean all work surfaces as directed by your teacher. Wash your hands thoroughly before leaving the lab.

Analysis

1. **Describing Events** Describe the size, shape, and numbers of spores you observed in the different fungus slides. Do you think you could identify a fungal species based on its spores alone? Explain.

____

____

____

____

____

2. **Examining Data** Compare the reproductive structures found in the sexually reproducing fungus cultures you observed. Do you think you could identify these fungal species based on their reproductive structures alone? Explain.

____

____

____

____

3. Classifying Suppose your teacher gives you an unknown fungus to identify. What questions would you ask to help you complete your task?

____

____

____

____

Conclusions

1. Making Predictions *Pilbous* is a member of the phylum Zygomycota. What would you expect the sexual reproductive structures of this fungus to look like?

____

____

2. Evaluating Methods In this lab, you learned that fungi are classified by their reproductive structures, and you observed the reproductive structures of representative fungi from each phylum. If you were to find an unknown fungus in a forest, what methods might you use to identify it?

____

____

____

____

____

____

Extensions

1. Designing Experiments Design an experiment to test the conditions under which a specific fungus will grow from spores. Then observe and describe the life cycle of the fungus you grow. Do not carry out your experiment without your teacher's permission.

2. Research and Communication Find out why some fungi that grow in and under homes are called toxic molds. What kinds of illnesses have been attributed to these fungi? What can be done to prevent these molds from growing in homes or to clean up homes that are already infected? How is the presence of these molds affecting the insurance industry?

Name ______________________ Class ____________ Date __________

Skills Practice Lab

CBL™ PROBEWARE

Best Food for Yeast

The flow of energy in almost all ecosystems begins with photosynthesis. During photosynthesis, plants, algae, and some bacteria make sugars by capturing the energy in sunlight. These organisms are called *producers*. They use the energy in sugars for their life processes. When producers are eaten by other organisms, called *consumers*, some of the energy in the producers is transferred to the consumers. Consumers use this energy for their life processes. One important group of consumers are the *decomposers*, which break down dead organisms into nutrients. Decomposers include yeasts, which are single-celled fungi.

As yeasts decompose organisms, some of the nutrients are returned to the environment, and some, such as sugars, are used by the yeasts for energy. One of the processes by which yeasts get energy from sugars is *fermentation*, which occurs in the absence of oxygen. The products of fermentations are carbon dioxide, alcohol, and energy. Enzymes in the yeasts speed up fermentation. The enzymes are specific for certain sugars.

People use the carbon dioxide produced during yeast fermentation to make bread rise and to make soda fizzy. When a baker uses yeast to make bread, the yeast feeds on the sugar in the dough. The carbon dioxide produced causes the dough to rise, while the alcohol evaporates when the dough is baked.

In this lab, you will determine the best sugar to use to maximize carbon dioxide production by yeast. You will use a CO_2 gas sensor to measure the amount of carbon dioxide produced during yeast fermentation.

OBJECTIVES

Use a CO_2 gas sensor to measure concentrations of carbon dioxide.

Compare three different sugars for their effect on carbon dioxide production by yeast during fermentation.

MATERIALS

- CBL System
- CO_2 gas sensor
- distilled water (2 mL)
- Erlenmeyer flasks, 250 mL (4)
- fructose solution, 5% (2 mL)
- pipet bulb
- pipets, 5 mL, labeled (5)
- lab apron
- lactose solution, 5% (2 mL)
- safety goggles
- sucrose solution, 5% (2 mL)
- test-tube rack
- test tubes, 10 mm × 100 mm (4)
- TI graphing calculator
- water bath
- wax pencil
- yeast suspension (8 mL)

Procedure

PREPARING YEAST CULTURES

1. Put on safety goggles and a lab apron.
2. Use a wax pencil to label four test tubes “water,” “fructose,” “lactose,” and “sucrose.” Write the same labels on four Erlenmeyer flasks. Place the test tubes in a test-tube rack. **CAUTION: If a piece of glassware breaks, do not pick it up with your bare hands. Notify your teacher immediately.**
3. Remove the bottle of distilled water from the water bath. Use the pipet labeled “distilled water” to transfer 2 mL of distilled water to the test tube labeled “water.” Return the bottle of distilled water to the water bath.
4. Repeat step 3 with the fructose, lactose, and sucrose solutions. Transfer each solution to the appropriately labeled test tube. Use the appropriately labeled pipet for each solution.
5. Remove the bottle of yeast suspension from the water bath. Gently swirl the bottle to mix the suspension evenly. Use the pipet labeled “yeast” to add 2 mL of yeast suspension to each test tube. Gently swirl each test tube. Return the bottle of yeast suspension to the water bath.
6. Place the test-tube rack containing the four test tubes in the water bath. Leave the test tubes in the water bath for at least 10 minutes while you complete steps 7–12.

SETTING UP THE CBL SYSTEM

7. Connect the CO_2 gas sensor to an adapter cable. Plug the adapter cable into the Channel 1 input of the CBL unit. Use the black cable to connect the CBL unit to the graphing calculator.
8. Turn on both the CBL unit and the calculator. Start the CHEMBIO program and go to the MAIN MENU.
9. Select SET UP PROBES. Enter “1” as the number of probes.
10. Select MORE PROBES from the SELECT PROBE menu. Select CO2 SENSOR from the SELECT PROBE menu. Enter “1” as the channel number. Select USE STORED from the CALIBRATION menu.
11. From the MAIN MENU, select COLLECT DATA. Select TIME GRAPH from the DATA COLLECTION menu. Enter “30” as the number of seconds between samples. Enter “10” as the number of samples. The CBL unit will record the carbon dioxide concentration twice each minute for 5 minutes. Press ENTER.

12. Select USE TIME SETUP from the CONTINUE screen. Enter "500" as the minimum concentration (Y min). Enter "5000" as the maximum concentration (Y max). Enter "500" as the concentration increment (Y scl). Do not start collecting data until step 15.

MEASURING CARBON DIOXIDE CONCENTRATIONS

13. Remove the test tube labeled "water" from the water bath. Empty the contents of the test tube into the flask labeled "water."

14. Immediately insert the CO_2 gas sensor into the flask by holding the rubber stopper and gently twisting it, as shown in **Figure 1.** Do not twist the shaft of the CO_2 gas sensor or you may damage it.

15. Press ENTER to begin measuring the carbon dioxide concentration in the flask. (If the calculator has turned off, turn it back on by pressing ON.) After 5 minutes, the CBL unit will stop recording.

FIGURE 1
INSERT GAS SENSOR

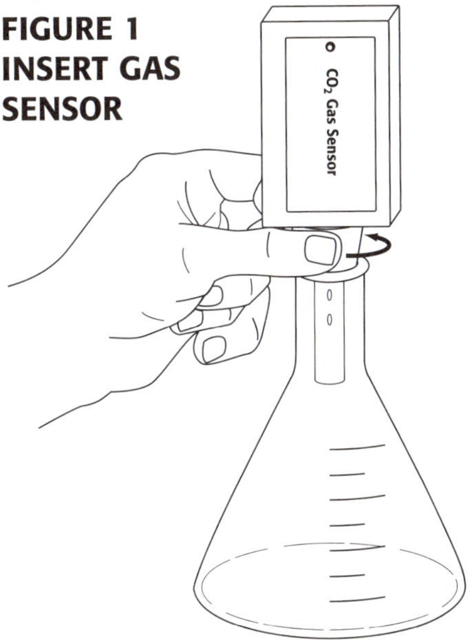

16. Press ENTER on the calculator to display a graph of carbon dioxide concentration (in parts per million, ppm) versus time (in seconds). Using the left and right arrow keys, scroll through the data points and record each carbon dioxide concentration in the "Water" column in **Table 1.**

17. Press ENTER again. Remove the CO_2 gas sensor from the flask.

18. Use your hand or a notebook to fan air across the openings in the shaft of the CO_2 gas sensor for 1 minute.

19. The REPEAT menu will be displayed on the calculator. Press YES to run another trial.

20. Repeat steps 13–19 with the fructose solution. Record the data in the "Fructose" column in **Table 1.**

21. Repeat steps 13–19 with the lactose solution. Record the data in the "Lactose" column in **Table 1.**

22. Repeat steps 13–19 with the sucrose solution. Record the data in the "Sucrose" column in **Table 1.**

23. Press NO in the REPEAT menu. Choose QUIT to exit the program. Turn both the CBL unit and the calculator off.

24. Put away your materials, and clean up your work area. Wash your hands before leaving the lab.

TABLE 1 CO_2 CONCENTRATIONS IN YEAST CULTURES

Time(s)	Water	Fructose	Lactose	Sucrose
30				
60				
90				
120				
150				
180				
210				
240				
270				
300				
Change in CO_2 concentration				

Analysis

1. **Constructing Graphs** Graph your data on the chart in **Figure 2.** Plot the data for all four yeast cultures, using a different color or different symbols for each data set. Label your graph clearly.

FIGURE 2 CO_2 CONCENTRATIONS IN YEAST CULTURES

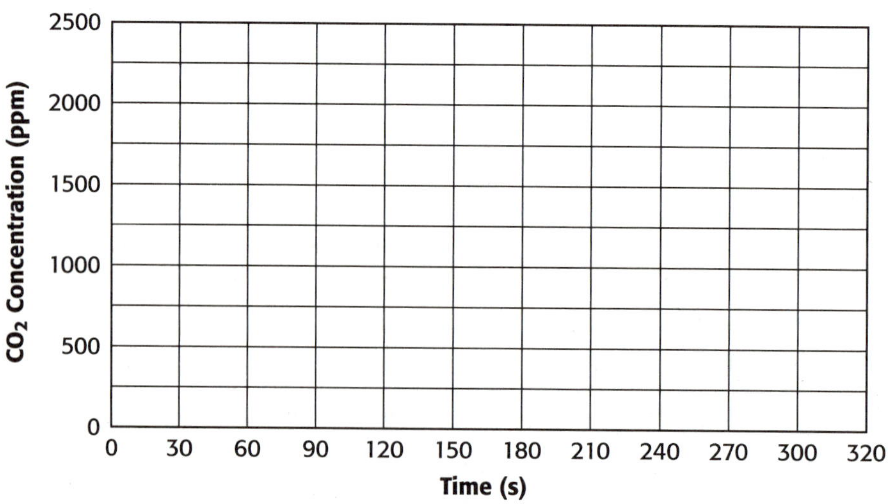

2. **Analyzing Graphs** Describe the changes in carbon dioxide concentration over time that you observed in the four yeast cultures.

____

____

____

____

3. **Organizing Data** Calculate the change in carbon dioxide concentration for all four yeast cultures by subtracting the initial concentration (at 0 seconds) from the final concentration (at 300 seconds) in each culture. Record the results in **Table 1.** If the final concentration is less than the initial concentration, enter "no change."
4. **Analyzing Results** Which culture had the largest change in carbon dioxide concentration? Which had the smallest?

____

____

5. **Analyzing Results** What is the relationship between fermentation and carbon dioxide concentration in this experiment? Rank the four cultures in order of increasing rate of fermentation.

____

____

____

____

____

Conclusions

1. **Interpreting Information** Suggest a possible reason why one yeast culture containing sugar may have a high rate of fermentation while another culture containing a different sugar has a low rate.

____

____

____

____

2. **Evaluating Methods** What is the purpose of having a culture that consists of yeast and distilled water?

___

___

___

3. **Evaluating Methods** How might the experiment be modified to show that yeasts are necessary for fermentation?

___

___

___

___

___

4. **Applying Conclusions** If you wanted to make a soda with a lot of carbonation (fizz), what would be the best yeast food (sugar) to use?

___

5. **Drawing Conclusions** In nature, what happens to the carbon dioxide produced by yeasts during fermentation?

___

___

___

Extensions

1. **Designing Experiments** Develop a procedure to find the preferred food of different kinds of yeast, such as dry yeast, cake yeast, and quick-starting yeast. Compare the fermentation rates of the different yeasts.
2. **Research and Communications** Investigate the characteristics of several species of yeasts found in nature. Identify the environments where they live and the food sources they normally use. Report your findings in an oral presentation.

Name ______________________ Class ______________ Date ____________

Quick Lab

DATASHEET FOR IN-TEXT LAB

Observing Bread Mold

You can use a microscope to see the individual threads of cells that make up the body of a fungus.

MATERIALS

- prepared slide of *Rhizopus*—black bread mold
- compound microscope

Procedure

1. Examine a slide of black bread mold under low power of a microscope.
2. Move the slide to an area where you can clearly see threadlike structures.
3. Draw what you see in your lab notebook. Be sure to use at least one-third of the page.
4. Move the slide to examine an area where you can clearly see the round bulb-like structures.
5. Draw what you see in your notebook.

Analysis

1. **Label** the drawings you made, including the hyphae and mycelium.

   Labels include hyphae, mycelium, and reproductive structures.

2. **Explain** where you would find each structure on a loaf of bread.

   The mycelium made of hyphae is throughout the bread. The reproductive structures are on the surface.

3. **Critical Thinking**
   Recognizing Relationships Relate the structures you drew to their functions described in your textbook.

   The hyphae absorb nutrients from the bread. The mycelium provides a high surface-to-volume ratio for nutrition. Reproductive structures release spores.

Name ______________________ Class ______________ Date ____________

Data Lab

DATASHEET FOR IN-TEXT LAB

Analyzing the Effect of Mycorrhizae

Background

Two groups of plants were planted in similar soils under similar conditions, but group A was grown in sterilized soil and group B was grown in nonsterilized soil. After 18 weeks of growth, a photograph was taken of the plants. Examine the drawing, and answer the following questions:

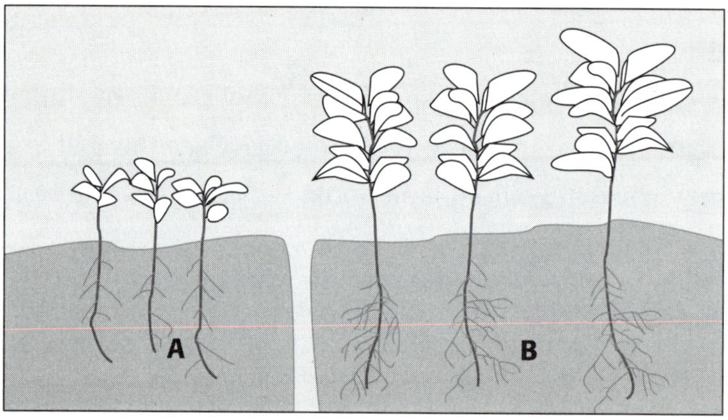

Analysis

1. **Compare** the growth of the two groups. Which grew faster?

   Group B

2. **Explain** why one group grew better than the other group.

   Sterilizing the soil kills all of the mycorrhizal fungi, which provide mineral nutrients to plants.

3. **Critical Thinking**
   Inferring Relationships Suggest a possible cause of slower growth in the smaller plants.

   Without the symbiotic fungi, the plants did not have as many nutrients available to them.

4. **Recommend** a course of action to restore growth in the stunted plants.

   Inoculate the soil of the stunted plants with mycorrhizal fungi.

Name ______________________________ Class ______________ Date ____________

Skills Practice Lab

DATASHEET FOR IN-TEXT LAB

Observing Yeast and Fermentation

SKILLS

- Observing
- Measuring
- Collecting data
- Analyzing data

OBJECTIVE

- **Observe** the release of energy by yeast during fermentation.

MATERIALS

- 500 mL vacuum bottle
- 10 cm glass tubing
- 2-hole rubber stopper
- 250 mL beaker
- 75 g sucrose
- one package dry yeast
- thermometer
- 50 cm rubber tubing
- 150 mL limewater

CHEMSAFETY

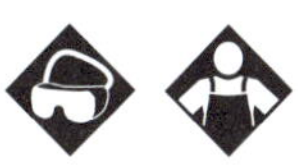

CAUTION: Always wear safety goggles and a lab apron to protect your eyes and clothing.

CAUTION: Glassware is fragile. Notify the teacher of broken glass or cuts. Do not clean up broken glass or spills with broken glass unless the teacher tells you to do so.

Before You Begin

Sucrose is a disaccharide—a carbohydrate made from two monosaccharides. It is one chemical made by plants to store the sun's energy. Yeast release the energy stored in sucrose in a process called **fermentation.** In this investigation you will have a chance to observe and measure the products of fermentation.

1. Write a definition for the boldface term in the paragraph above.

Fermentation is the process by which cells release the energy in sugar without using oxygen.

2. You will be using the data table below to record your data.

Data Table								
Fermentation by Yeast								
Time	**Date**	**Temp.**	**Time**	**Date**	**Temp.**	**Time**	**Date**	**Temp.**
1.			8.			15.		
2.			9.			16.		
3.			10.			17.		
4.			11.			18.		
5.			12.			19.		
6.			13.			20.		
7.			14.			21.		

Procedure

1. Set up your vacuum bottle according to the diagram.

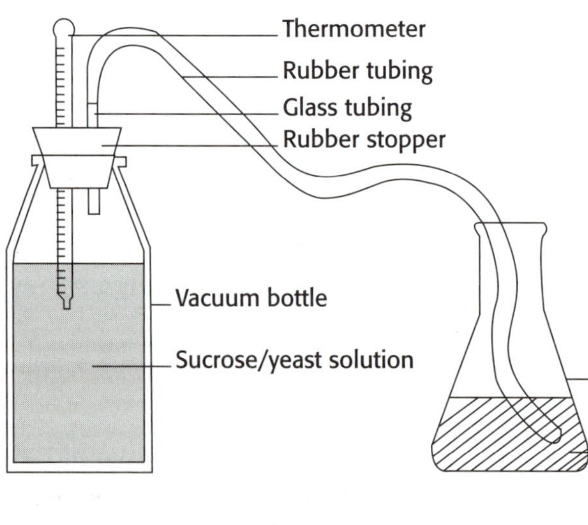

2. Mix 75 g of sucrose in 400 mL of water.

3. When the sucrose has dissolved, add one-half package of fresh yeast and stir.

4. Pour the sucrose-yeast solution into a vacuum bottle until it is approximately three-quarters full.

5. Adjust the thermometer so that it extends down into the sugar-yeast solution.

6. Record the temperature of the solution on the observation chart as soon as possible. Continue to record the temperature as often as possible during the next two days.

Analyze and Conclude

1. Summarizing Results Prepare a graph of your data, illustrating the temperature over time. Complete the graph by drawing a curve through the plotted points.

Students' graphs may vary. They should show that as fermentation progresses, the temperature in the sucrose-yeast solution increases.

2. Analyzing Data What does the curved line plotted on the graph indicate?

Answers may vary. Students should interpret the curved line as an indication that fermentation activity increases over time.

3. Drawing Conclusions What can you conclude about the energy contained in sucrose?

Answers may vary. Since the temperature of the sucrose-yeast solution increases over time, students should conclude that as fermentation occurs, large amounts of energy are being released. This would indicate that the sucrose molecules contain a large amount of energy in their molecular bonds, which is released as the yeast metabolizes the sucrose.

4. Predicting Patterns What do you think would happen if there were only one hole in the stopper for the thermometer?

Answers may vary. The carbon dioxide gas produced by the yeast would increase the pressure in the bottle, and the stopper would eventually come off.

5. Further Inquiry If you know that fermentation liberates energy and gives off carbon dioxide and alcohol as waste products, how would you prove that fermentation is really taking place in the sugar-yeast solution?

Answers may vary. Students could test the sucrose-yeast solution at the beginning of the experiment for the presence of carbon dioxide by adding a small amount to an indicator solution.

Exploration Lab

OBSERVATION

Exploring Fungi

Teacher Notes

TIME REQUIRED One 45-minute period

SKILLS ACQUIRED

Classifying
Collecting data
Identifying patterns
Inferring
Interpreting
Organizing and analyzing data

RATINGS Easy ← 1 2 3 4 → Hard

Teacher Prep–2
Student Setup–2
Concept Level–1
Cleanup–2

THE SCIENTIFIC METHOD

Make Observations Students make observations of fungi reproductive structures.

Ask Questions Analysis question 3 requires students to ask questions about how to identify an unknown fungus.

Analyze the Results Students analyze their observations in Analysis question 2.

MATERIALS

Provide students with prepared slides from each of the four fungal phyla. Examples include Zygomycota: *Rhizopus*; Basidiomycota: mushrooms, such as *Agaricus campestris* (which can be purchased at a local grocery store); Ascomycota: *Neurospora*; Deuteromycota: *Apergillus* and *Penicillium.*

Materials for this lab can be purchased from WARD'S. See the *Master Materials List* for ordering instructions.

SAFETY CAUTIONS

- Discuss all safety symbols and caution statements with students.
- Instruct students to follow the proper procedures for carrying and using the compound light microscope. Do not allow students to use microscopes with frayed or damaged electrical cords. Students should never use sunlight as a light source with microscopes that lack their own light source.

TECHNIQUES TO DEMONSTRATE

Help students locate the reproductive areas of each fungus.

Before allowing students to use a compound light microscope, demonstrate the proper way to carry, handle, and use it.

Demonstrate the proper way to view a slide. Raise the objectives (or lower the stage) as far as possible by turning the coarse-adjustment knob toward you. Secure a prepared slide to the stage, using the stage clips. Turn the low-power objective into position over the stage. While observing the stage from eye level, use the coarse-adjustment knob to position the objective as close to the slide as it will go without touching the slide. Demonstrate how to focus with the coarse-adjustment knob only by turning it toward you. Tell students never to focus by moving the objective downward. They may run the objective into the slide and damage the slide or the objective.

TIPS AND TRICKS

This lab works best in groups of two students.

Tell students to keep both eyes open while looking through the eyepiece of the microscope. This avoids eye strain.

Name ______________________ Class ______________ Date ____________

Exploration Lab

OBSERVATION

Exploring Fungi

Fungi are classified into four phyla based on their sexual reproductive structures. There are three phyla of sexually reproducing fungi: Zygomycota, Ascomycota, and Basidiomycota. Each group is characterized by distinctive sexual reproductive structures.

FIGURE 1 THREE PHYLA OF FUNGI

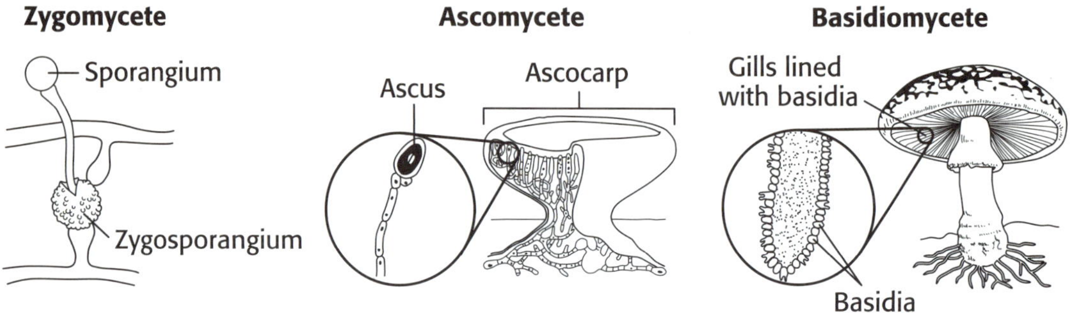

Zygomycetes form thick-walled sexual structures called *zygosporangia*. You may have seen zygomycetes growing on old bread in the form of mold. *Ascomycetes*, which include yeasts and truffles, can reproduce sexually and asexually. They produce sexual spores in the form of saclike *asci*. Basidiomycetes, which include mushrooms, form club-shaped *basidia* as sexual spores. The phylum called Deuteromycota includes more than 17,000 fungi, which appear to reproduce only asexually.

In this lab, you will examine and compare specimens from the four fungi phyla.

OBJECTIVES

Observe a specimen from each phylum of fungus.

Draw and label pictures of a representative sample from each phylym of fungus.

Identify where each type of fungus might be found and what it uses as a food source.

MATERIALS

- compound light microscope
- field guide to fungi
- lab apron
- representative slides of each phylum of fungi
- safety goggles
- transparent tape

Name ______________________ Class ______________ Date ____________

Exploring Fungi *continued*

Procedure

1. Choose one of the slides provided by your teacher. Record your observations about the fungi appearance in **Table 1.**

TABLE 1 OBSERVATIONS OF FOUR PHYLA OF FUNGI

Phylum and specimen name	Appearance	Drawing of reproductive structures	Location and food sources
Zygomycota Culture: ______________			
Ascomycota Culture: ______________			
Basidiomycota Culture: ______________			
Deuteromycota Culture: ______________			

2. Locate the reproductive area of the fungus. If you cannot find the reproductive structures, look in your textbook or ask your teacher for help.

Name ______________________ Class ______________ Date ______________

3. Observe the reproductive structures by using the low power of the microscope. Record your observations by drawing and labeling what you see. If your drawing will not fit in the space provided in **Table 1,** then use a separate sheet of paper.
4. Use your textbook, a field guide, or other resource to determine where this type of fungus might be found. What are some possible food sources for this type of fungus? Record your answers in **Table 1.**
5. Repeat steps 1–4 for each slide provided.
6. Dispose of your materials according to your teacher's instructions.
7. Clean all work surfaces as directed by your teacher. Wash your hands thoroughly before leaving the lab.

Analysis

1. **Describing Events** Describe the size, shape, and numbers of spores you observed in the different fungus slides. Do you think you could identify a fungal species based on its spores alone? Explain.

   Answers will vary and should be based on students' data. Most students will note that the spores they observed were microscopic and round and that they were produced in large numbers. Students could probably not identify a fungal species based only on spores because the spores look very similar to one another.

2. **Examining Data** Compare the reproductive structures found in the sexually reproducing fungus cultures you observed. Do you think you could identify these fungal species based on their reproductive structures alone? Explain.

   Answers will vary and should be based on students' data. Most students will note that zygosporangia are thick-walled, asci are saclike, and basidia are club-shaped. Students should be able to identify each culture by its reproductive structures because the structures are distinct.

Name ______________________ Class ______________ Date ______________

Exploring Fungi *continued*

3. **Classifying** Suppose your teacher gives you an unknown fungus to identify. What questions would you ask to help you complete your task?

Answers will vary. Students should realize that fungi are classified by their reproductive structures. Possible questions include: How does the fungus reproduce? What is its life cycle? What do its sexual reproductive structures look like?

Conclusions

1. **Making Predictions** *Pilbous* is a member of the phylum Zygomycota. What would you expect the sexual reproductive structures of this fungus to look like?

Answers will vary. Students might predict that the sexual reproductive structures will be thick-walled and similar to those of other members of this phylum.

2. **Evaluating Methods** In this lab, you learned that fungi are classified by their reproductive structures, and you observed the reproductive structures of representative fungi from each phylum. If you were to find an unknown fungus in a forest, what methods might you use to identify it?

Answers will vary. Students should realize that an unknown fungus in nature can be dangerous and that locating the reproductive structures is not always easy. Students might suggest using a field guide to identify the fungus instead of observing the reproductive structures under a microscope.

Extensions

1. **Designing Experiments** Design an experiment to test the conditions under which a specific fungus will grow from spores. Then observe and describe the life cycle of the fungus you grow. Do not carry out your experiment without your teacher's permission.
2. **Research and Communication** Find out why some fungi that grow in and under homes are called toxic molds. What kinds of illnesses have been attributed to these fungi? What can be done to prevent these molds from growing in homes or to clean up homes that are already infected? How is the presence of these molds affecting the insurance industry?

Skills Practice Lab

CBL™ PROBEWARE

Best Food for Yeast

Teacher Notes

TIME REQUIRED One 45-minute period

SKILLS ACQUIRED

Collecting data
Experimenting
Identifying patterns
Inferring
Interpreting
Measuring
Organizing and analyzing data

RATINGS

Easy ← 1 2 3 4 → Hard

Teacher Prep–3
Student Setup–3
Concept Level–2
Cleanup–2

THE SCIENTIFIC METHOD

Make Observations Students observe changes in carbon dioxide concentration over time in cultures containing yeast and one of three kinds of sugars.

Analyze the Results Analysis questions 2, 4, and 5 require students to analyze their results.

Draw Conclusions Conclusions questions 4 and 5 ask students to draw conclusions from their data.

MATERIALS

Materials for this lab can be purchased from WARD'S. See the *Master Materials List* for ordering instructions. Yeast and sucrose (table sugar) can be purchased at a supermarket. Glucose may be substituted for fructose or sucrose.

SAFETY CAUTIONS

- Discuss all safety symbols and caution statements with students.
- Caution students not to taste the sugar solutions or the yeast suspension.

DISPOSAL

Store-bought dry yeast can be mixed with water and poured down the drain. All other dry yeast should be destroyed by autoclaving or disinfecting with household bleach (5% sodium hypochlorite). Yeast suspensions and solutions can be rinsed down the drain under running water.

TECHNIQUES TO DEMONSTRATE

If you wish to have students use a computer printer to print the graphs on their calculators, demonstrate how to use the TI-Graph link cable and TI Connect software.

TIPS AND TRICKS

Preparation

This lab works best in groups of two to four students.

Prepare the 5% lactose, sucrose, and fructose solutions by dissolving 5 g of each sugar in 100 mL of distilled water. Prepare the yeast suspension by dissolving one 1/4 oz (7 g) package of yeast in 100 mL of distilled water. These volumes will be enough for approximately 30–40 groups.

At least 15 minutes before the start of class, place bottles or flasks containing distilled water, the yeast suspension, and the three sugar solutions in a water bath set at 30–35°C. The bottles or flasks should be heavy enough to remain upright in the water bath.

For each group of students, provide five 5 mL pipets labeled "distilled water," "yeast," "fructose," "lactose," and "sucrose," respectively.

The procedure in this lab is written for use with the original CBL system. If you are using CBL 2 or LabPro, the CHEMBIO program can still be used. Updated versions of this program can be downloaded from **www.vernier.com**. For additional information on how to integrate the CBL system into your laboratory, see the Program Introduction.

Some sensors may require the use of an adapter. Students will need to connect the adapter to the sensor before connecting it to the CBL.

Procedure

Procedure step 12 calls for students to set up the calculator to collect data up to 5,000 ppm in increments of 500 ppm. If the maximum carbon dioxide concentrations are much smaller or larger than 5,000 ppm, have students increase or decrease the maximum concentration (Y max) and concentration increment (Y scl).

Remind students to twist the rubber stopper, not the shaft of the CO_2 gas sensor, when they insert the sensor into each flask. Twisting the shaft may damage the sensor.

Caution students not to allow any air to enter the flasks while the CBL System is collecting data.

Tell students that the CO_2 gas sensor measures the concentration of carbon dioxide that has diffused into the shaft. Explain that they need to fan air across the openings in the shaft between trials to disperse gases that remain in the shaft from the previous trial.

Name ______________________________ Class ______________ Date ______________

Skills Practice Lab

CBL™ PROBEWARE

Best Food for Yeast

The flow of energy in almost all ecosystems begins with photosynthesis. During photosynthesis, plants, algae, and some bacteria make sugars by capturing the energy in sunlight. These organisms are called *producers*. They use the energy in sugars for their life processes. When producers are eaten by other organisms, called *consumers*, some of the energy in the producers is transferred to the consumers. Consumers use this energy for their life processes. One important group of consumers are the *decomposers*, which break down dead organisms into nutrients. Decomposers include yeasts, which are single-celled fungi.

As yeasts decompose organisms, some of the nutrients are returned to the environment, and some, such as sugars, are used by the yeasts for energy. One of the processes by which yeasts get energy from sugars is *fermentation*, which occurs in the absence of oxygen. The products of fermentations are carbon dioxide, alcohol, and energy. Enzymes in the yeasts speed up fermentation. The enzymes are specific for certain sugars.

People use the carbon dioxide produced during yeast fermentation to make bread rise and to make soda fizzy. When a baker uses yeast to make bread, the yeast feeds on the sugar in the dough. The carbon dioxide produced causes the dough to rise, while the alcohol evaporates when the dough is baked.

In this lab, you will determine the best sugar to use to maximize carbon dioxide production by yeast. You will use a CO_2 gas sensor to measure the amount of carbon dioxide produced during yeast fermentation.

OBJECTIVES

Use a CO_2 gas sensor to measure concentrations of carbon dioxide.

Compare three different sugars for their effect on carbon dioxide production by yeast during fermentation.

MATERIALS

- CBL System
- CO_2 gas sensor
- distilled water (2 mL)
- Erlenmeyer flasks, 250 mL (4)
- fructose solution, 5% (2 mL)
- pipet bulb
- pipets, 5 mL, labeled (5)
- lab apron
- lactose solution, 5% (2 mL)
- safety goggles
- sucrose solution, 5% (2 mL)
- test-tube rack
- test tubes, 10 mm × 100 mm (4)
- TI graphing calculator
- water bath
- wax pencil
- yeast suspension (8 mL)

Name ______________________ Class ______________ Date ____________

Procedure

PREPARING YEAST CULTURES

1. Put on safety goggles and a lab apron.
2. Use a wax pencil to label four test tubes "water," "fructose," "lactose," and "sucrose." Write the same labels on four Erlenmeyer flasks. Place the test tubes in a test-tube rack. **CAUTION: If a piece of glassware breaks, do not pick it up with your bare hands. Notify your teacher immediately.**
3. Remove the bottle of distilled water from the water bath. Use the pipet labeled "distilled water" to transfer 2 mL of distilled water to the test tube labeled "water." Return the bottle of distilled water to the water bath.
4. Repeat step 3 with the fructose, lactose, and sucrose solutions. Transfer each solution to the appropriately labeled test tube. Use the appropriately labeled pipet for each solution.
5. Remove the bottle of yeast suspension from the water bath. Gently swirl the bottle to mix the suspension evenly. Use the pipet labeled "yeast" to add 2 mL of yeast suspension to each test tube. Gently swirl each test tube. Return the bottle of yeast suspension to the water bath.
6. Place the test-tube rack containing the four test tubes in the water bath. Leave the test tubes in the water bath for at least 10 minutes while you complete steps 7–12.

SETTING UP THE CBL SYSTEM

7. Connect the CO_2 gas sensor to an adapter cable. Plug the adapter cable into the Channel 1 input of the CBL unit. Use the black cable to connect the CBL unit to the graphing calculator.
8. Turn on both the CBL unit and the calculator. Start the CHEMBIO program and go to the MAIN MENU.
9. Select SET UP PROBES. Enter "1" as the number of probes.
10. Select MORE PROBES from the SELECT PROBE menu. Select CO2 SENSOR from the SELECT PROBE menu. Enter "1" as the channel number. Select USE STORED from the CALIBRATION menu.
11. From the MAIN MENU, select COLLECT DATA. Select TIME GRAPH from the DATA COLLECTION menu. Enter "30" as the number of seconds between samples. Enter "10" as the number of samples. The CBL unit will record the carbon dioxide concentration twice each minute for 5 minutes. Press ENTER.

Name ______________________ Class ______________ Date ____________

12. Select USE TIME SETUP from the CONTINUE screen. Enter "500" as the minimum concentration (Y min). Enter "5000" as the maximum concentration (Y max). Enter "500" as the concentration increment (Y scl). Do not start collecting data until step 15.

MEASURING CARBON DIOXIDE CONCENTRATIONS

13. Remove the test tube labeled "water" from the water bath. Empty the contents of the test tube into the flask labeled "water."

14. Immediately insert the CO_2 gas sensor into the flask by holding the rubber stopper and gently twisting it, as shown in **Figure 1.** Do not twist the shaft of the CO_2 gas sensor or you may damage it.

15. Press ENTER to begin measuring the carbon dioxide concentration in the flask. (If the calculator has turned off, turn it back on by pressing ON.) After 5 minutes, the CBL unit will stop recording.

FIGURE 1
INSERT GAS SENSOR

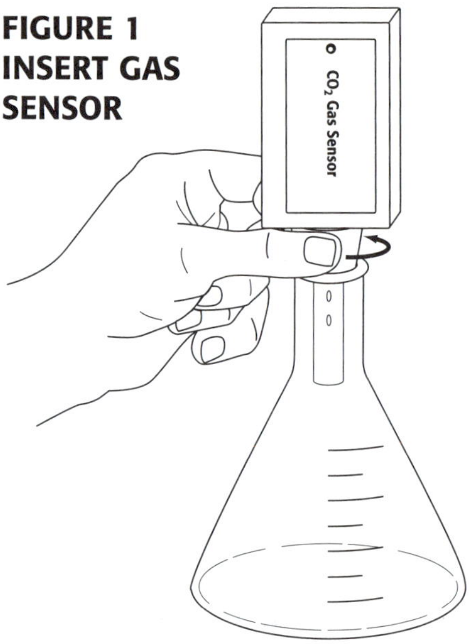

16. Press ENTER on the calculator to display a graph of carbon dioxide concentration (in parts per million, ppm) versus time (in seconds). Using the left and right arrow keys, scroll through the data points and record each carbon dioxide concentration in the "Water" column in **Table 1.**

17. Press ENTER again. Remove the CO_2 gas sensor from the flask.

18. Use your hand or a notebook to fan air across the openings in the shaft of the CO_2 gas sensor for 1 minute.

19. The REPEAT menu will be displayed on the calculator. Press YES to run another trial.

20. Repeat steps 13–19 with the fructose solution. Record the data in the "Fructose" column in **Table 1.**

21. Repeat steps 13–19 with the lactose solution. Record the data in the "Lactose" column in **Table 1.**

22. Repeat steps 13–19 with the sucrose solution. Record the data in the "Sucrose" column in **Table 1.**

23. Press NO in the REPEAT menu. Choose QUIT to exit the program. Turn both the CBL unit and the calculator off.

24. Put away your materials, and clean up your work area. Wash your hands before leaving the lab.

Name ______________________ Class ______________ Date ____________

Best Food for Yeast *continued*

TABLE 1 CO_2 CONCENTRATIONS IN YEAST CULTURES

Time(s)	Water	Fructose	Lactose	Sucrose
30	**560**	**963**	**828**	**1,276**
60	**559**	**1,018**	**884**	**1,478**
90	**593**	**1,097**	**940**	**1,690**
120	**616**	**1,164**	**973**	**1,825**
150	**661**	**1,276**	**1,108**	**1,926**
180	**650**	**1,343**	**1,175**	**2,004**
210	**640**	**1,432**	**1,231**	**2,094**
240	**654**	**1,500**	**1,287**	**2,160**
270	**655**	**1,567**	**1,309**	**2,228**
300	**658**	**1,635**	**1,332**	**2,385**
Change in CO_2 concentration	**98**	**672**	**504**	**1,109**

Entries will vary for each group. Sample data are entered above.

Analysis

1. **Constructing Graphs** Graph your data on the chart in **Figure 2.** Plot the data for all four yeast cultures, using a different color or different symbols for each data set. Label your graph clearly.

FIGURE 2 CO_2 CONCENTRATIONS IN YEAST CULTURES

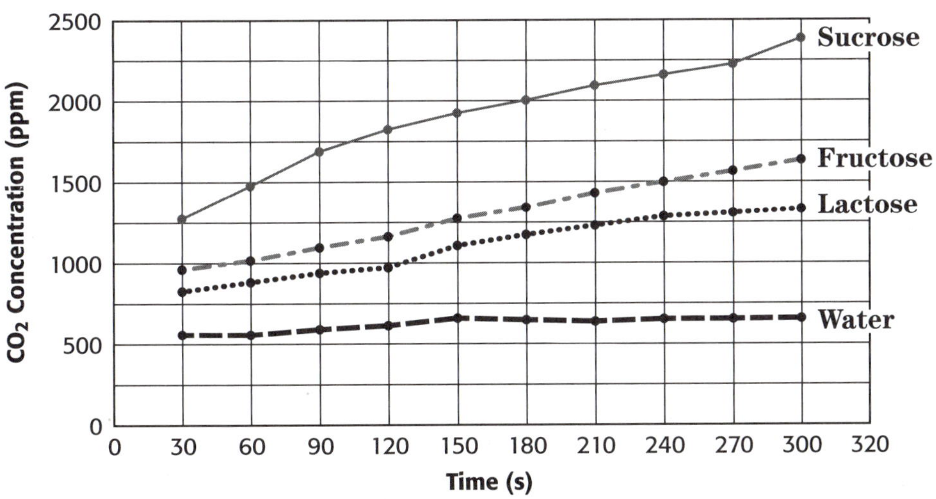

Graphs will vary, based on each group's data. The graph shown is based on the sample data given in Table 1.

2. **Analyzing Graphs** Describe the changes in carbon dioxide concentration over time that you observed in the four yeast cultures.

Answers will vary. The fructose and sucrose cultures should show a rapid increase in carbon dioxide concentration over time. The lactose culture should show a more gradual increase. There should be little or no change in carbon dioxide concentration in the water culture.

3. **Organizing Data** Calculate the change in carbon dioxide concentration for all four yeast cultures by subtracting the initial concentration (at 0 seconds) from the final concentration (at 300 seconds) in each culture. Record the results in **Table 1.** If the final concentration is less than the initial concentration, enter "no change."

4. **Analyzing Results** Which culture had the largest change in carbon dioxide concentration? Which had the smallest?

Answers will vary. Either the fructose or the sucrose culture should show the largest change. The water culture should show the smallest change.

5. **Analyzing Results** What is the relationship between fermentation and carbon dioxide concentration in this experiment? Rank the four cultures in order of increasing rate of fermentation.

As fermentation takes place in a flask, the carbon dioxide concentration in the flask increases. Rankings will vary. Water should be the first culture listed (lowest rate), and lactose should be the second. Third and fourth should be fructose and sucrose, in whichever order agrees with the data that were collected.

Conclusions

1. **Interpreting Information** Suggest a possible reason why one yeast culture containing sugar may have a high rate of fermentation while another culture containing a different sugar has a low rate.

Accept any plausible explanation. One possibility is that yeast may be unable to take up certain sugars from their surroundings. Another is that yeast may lack the enzymes needed to catalyze the fermentation of certain sugars.

Name ______________________ Class ______________ Date ____________

2. **Evaluating Methods** What is the purpose of having a culture that consists of yeast and distilled water?

This culture serves as a control against which the data from the cultures containing sugars can be compared. It shows that fermentation cannot take place unless a sugar is present.

3. **Evaluating Methods** How might the experiment be modified to show that yeasts are necessary for fermentation?

Perform the same procedure on three flasks that contain fructose, lactose, and sucrose with yeast and three flasks that contain the same concentrations of these sugars without yeast. To keep the sugar concentrations the same, add 2 mL of distilled water to the last three flasks instead of yeast.

4. **Applying Conclusions** If you wanted to make a soda with a lot of carbonation (fizz), what would be the best yeast food (sugar) to use?

Fructose or sucrose would be the best food for yeast.

5. **Drawing Conclusions** In nature, what happens to the carbon dioxide produced by yeasts during fermentation?

It is released into the soil, water, and air, thereby becoming part of the carbon cycle.

Extensions

1. **Designing Experiments** Develop a procedure to find the preferred food of different kinds of yeast, such as dry yeast, cake yeast, and quick-starting yeast. Compare the fermentation rates of the different yeasts.
2. **Research and Communications** Investigate the characteristics of several species of yeasts found in nature. Identify the environments where they live and the food sources they normally use. Report your findings in an oral presentation.

Answer Key

Directed Reading

SECTION: CHARACTERISTICS OF FUNGI

1. heterotrophic
2. chitin
3. The bodies of fungi are made up of long, slender filaments that are woven together.
4. Nuclear mitosis is the type of mitosis that occurs in fungi. The nuclear envelope remains intact during nuclear mitosis, spindle fibers form within the nucleus, and the nuclear envelope pinches the nucleus in two.
5. Hyphae are the slender filaments that make up the bodies of fungi.
6. A mycelium is a tangled mass of hyphae formed when hyphae grow.
7. All fungi secrete digestive enzymes that break down organic matter in their environment. The decomposed molecules are then absorbed.
8. spores
9. haploid
10. asexual reproduction
11. mating types

SECTION: FUNGAL DIVERSITY

1. Deuteromycetes are fungi in which no sexual stage has been seen. Most are now classified in phylum Ascomycota.
2. Some deuteromycetes are the fungi that cause ringworm and athlete's foot, *Penicillium* species that produce antibiotics and are used in making cheese, and *Aspergillis* species that are used to ferment soy sauce and produce citric acid.
3. Truffles are in phylum Ascomycota. Members of this phylum produce sexual spores in asci and have dividing walls in their hyphae.
4. The hyphae in Ascomycota are divided by walls, and the hyphae in Zygomycota are not.
5. Mushrooms are in the phylum Basidiomycota. Members of this phylum produce sexual spores in basidia and have dividing walls in their hyphae.
6. c
7. a
8. b
9. ascus
10. yeast
11. budding
12. sexually
13. reproductive
14. asexually
15. poisonous

SECTION: FUNGAL PARTNERSHIPS

1. A mycorrhiza is a mutualistic relationship formed between fungi and the roots of vascular plants.
2. The fungi obtain carbohydrates made by the plants.
3. The plants are able to get more mineral nutrients, which the fungi absorb from the soil and transfer to the plants' roots.
4. penetrate
5. basidiomycete
6. A lichen is a symbiosis that consists of a fungus and a photosynthetic partner, such as a green alga or a cyanobacterium.
7. The fungus protects the photosynthetic partner from the environment and provides it with mineral nutrients.
8. The tough construction of the fungus, the photosynthetic ability of the alga or cyanobacterium, and the ability of the lichen to become dormant during drought and freezing help lichens to survive in harsh habitats.

Active Reading

SECTION: CHARACTERISTICS OF FUNGI

1. because both groups of organisms are immobile, have a cell wall, and appear "rooted" in the soil
2. A mushroom, like other fungi, lacks chlorophyll. Plants contain this pigment, which gives a plant its green coloring.
3. by breaking down organic molecules that they absorb from their environment

4. In most eukaryotes, the nuclear envelope disintegrates in prophase and reforms in telophase. In fungi, the nuclear envelope remains intact from prophase to anaphase.
5. f
6. f
7. b
8. p
9. f
10. b
11. p
12. p
13. b
14. f
15. b

SECTION: FUNGAL DIVERSITY

1. c
2. e
3. a
4. d
5. b
6. b

SECTION: FUNGAL PARTNERSHIPS

1. The photosynthetic partner can carry out photosynthesis.
2. The fungus has a sturdy structure, and the alga can carry out photosynthesis.
3. d

Vocabulary Review

1. c
2. f
3. g
4. b
5. h
6. e
7. i
8. j
9. l
10. k
11. a
12. d

Science Skills

CLASSIFYING

Oak tree

eukaryotic
photosynthetic
has cell walls
appears "rooted" in the soil
has roots
contains chlorophyll
contains cellulose
many cell types and tissue types
immobile

Mushroom

eukaryotic
heterotrophic
has cell walls
appears "rooted" in the soil
has rhizoids
absorbs organic nutrients from the environment
contains chitin
has hyphae
body of long, slender filaments
immobile

Both oak tree and mushroom

eukaryotic
has cell walls
appears "rooted" in the soil
immobile

Concept Mapping

1. sexually
2. asexually
3. zygomycetes
4. ascomycetes
5. mushrooms or puffballs
6. puffballs or mushrooms
7. truffles or morels
8. morels or truffles
9. deuteromycetes

Critical Thinking

1. c
2. b
3. d
4. a
5. d
6. c
7. b
8. e
9. a
10. a
11. c
12. b
13. e, h
14. a, b
15. c, f
16. g, d
17. a
18. d
19. a
20. d

Test Prep Pretest

1. c
2. b
3. d
4. b
5. d
6. c
7. chitin
8. hyphae
9. stolons
10. lichens
11. carbohydrates
12. fungus, ascomycete
13. digestive enzymes
14. hyphae

15. zygosporangia, asci, basidia
16. decompose
17. mycorrhizae
18. gills, basidia
19. parasitic
20. yeasts
21. Fungi do not contain chlorophyll, and they are heterotrophic. They have filamentous bodies instead of a variety of cell and tissue types. Their cell walls are made of chitin. They undergo nuclear mitosis.
22. In plants and most eukaryotes, the nuclear envelope breaks down in prophase and re-forms during telophase. In dividing mushroom cells, the nuclear envelope remains whole from prophase to anaphase. This results in spindle fibers forming within the nucleus. The spindle fibers drag the chromosomes to opposite poles of the nucleus instead of to opposite poles of the cell. Mitosis is completed when the nuclear envelope pinches in two.
23. The tough construction of the fungus, combined with the photosynthetic abilities of the alga or cyanobacterium, has enabled lichens to live in the harshest of habitats. They can survive drought and freezing by becoming dormant.
24. Beneficial ascomycetes include morels, truffles, and baker's yeast. Harmful ascomycetes include *Endothia parasitica*, which causes chestnut blight, and yeasts that cause thrush.
25. *Amanita muscaria*, like all fungi, is a heterotroph. It secretes enzymes that break down organic materials into simple molecules that the hyphae can absorb.

Quiz

SECTION: CHARACTERISTICS OF FUNGI

1. b	6. c
2. a	7. c
3. c	8. d
4. b	9. c
5. a	10. a

SECTION: FUNGAL DIVERSITY

1. e	6. c
2. d	7. d
3. f	8. a
4. a	9. c
5. b	10. d

SECTION: FUNGAL PARTNERSHIPS

1. c	6. d
2. c	7. b
3. a	8. d
4. c	9. b
5. a	10. d

Chapter Test (General)

1. a	11. b
2. b	12. d
3. c	13. b
4. b	14. a
5. c	15. c
6. d	16. c
7. c	17. b
8. d	18. d
9. b	19. d
10. c	20. b

Chapter Test (Advanced)

1. d	7. a
2. b	8. c
3. d	9. b
4. b	10. c
5. a	11. d
6. d	

12. organic matter
13. spore
14. ascus
15. mycelium, hyphae
16. zygosporangia, asci, basidia
17. infections or diseases
18. nuclear
19. yeasts
20. deuteromycetes
21. lichens
22. A mycorrhiza is a relationship between a fungus and the roots of vascular plants. The fungal hyphae can either penetrate the outer cells of the root or wrap around the root. The fungus transfers minerals from the soil to the plant. The plant provides the fungus with carbohydrates. A lichen is a relationship between a fungus and a

photosynthetic partner such as an alga, a cyanobacterium, or both. The fungus, usually an ascomycete, protects the photosynthetic partner and provides it with mineral nutrients. The photosynthetic partner provides the fungus with carbohydrates.

23. The mycelium of a fungus can be made of many meters of individual hyphae. This creates a high surface area-to-volume ratio, which makes a fungus well suited for absorbing food from the environment.

24. In asexual reproduction in the zygomycetes, haploid spores are produced on the tips of specialized hyphae. At maturity, the spores are shed and carried by the wind to new locations, where they germinate.

25. Fossil evidence indicates that the earliest land plants often had mycorrhizae. The soil at that time lacked organic matter. Plants with mycorrhizae are particularly successful in infertile soil.

Lesson Plan

Section: Characteristics of Fungi

Pacing

Regular Schedule:	**with lab(s):** N/A	**without lab(s):** 3 days
Block Schedule:	**with lab(s):** N/A	**without lab(s):** 1 1/2 days

Objectives

1. List the characteristics of kingdom Fungi.
2. Describe the structure of a typical fungus body.
3. Identify how fungi obtain nutrients.
4. Relate the way fungi obtain nutrients to their role in ecosystems.
5. Distinguish the ways that fungi reproduce.

National Science Education Standards Covered

UNIFYING CONCEPTS AND PROCESSES

UCP5: Form and function

SCIENCE AS INQUIRY

SI1: Abilities necessary to do scientific inquiry

SI2: Understandings about scientific inquiry

LIFE SCIENCE: THE CELL

LSCell1: Cells have particular structures that underlie their functions.

LSCell2: Most cell functions involve chemical reactions.

LIFE SCIENCE: THE MOLECULAR BASIS OF HEREDITY

LSGene1: In all organisms, the instructions for specifying the characteristics of the organisms are carried in DNA.

LIFE SCIENCE: BIOLOGICAL EVOLUTION

LSEvol5: Biological classifications are based on how organisms are related.

LIFE SCIENCE: INTERDEPENDENCE OF ORGANISMS

LSInter3: Organisms both cooperate and compete in ecosystems.

LSInter5: Human beings live within the world's ecosystems.

KEY
SE = Student Edition **TE** = Teacher Edition
CRF = Chapter Resource File

Block 1

CHAPTER OPENER *(45 minutes)*

_ **Quick Review,** SE. Students answer questions covered in previous sections of the textbook as preparation for the chapter content. **(GENERAL)**

_ **Reading Activity,** SE. Students agree or disagree with three statements about fungi. After reading the chapter, students return to the questions to see if their opinions have changed. **(GENERAL)**

_ **Using the Figure,** TE. Students answer questions about the chapter opener photograph. **(GENERAL)**

_ **Opening Activity**, Identifying Fungi, TE. Students ask yes or no questions to identify an object in a bag (a mushroom). **(GENERAL)**

Block 2

FOCUS *(5 minutes)*

_ **Bellringer Transparency.** Use this transparency as students enter the classroom and find their seats. **(GENERAL)**

MOTIVATE *(10 minutes)*

_ **Activity**, TE. Students prepare two slices of bread—one damp, one dry—in two separate, sealed plastic bags to see which will grow mold. Discuss what is required for mold growth. **(BASIC)**

TEACH *(30 minutes)*

_ **Teaching Transparency, Section Outline.** Use this transparency to give students a framework for the information in this section. **(GENERAL)**

_ **Inclusion Strategies**, TE. Students list harmful effects and beneficial uses of different kinds of fungi. **(GENERAL)**

_ **Teaching Tip,** Comparing Fungi and Plants, TE. Students use key words written on index cards to discuss how plants and fungi are alike and different. **(GENERAL)**

_ **Using the Figure,** Figure 2, TE. Students compare a moldy orange cut in half with the figure. **(GENERAL)**

HOMEWORK

_ **Active Reading Worksheet, Characteristics of Fungi, CRF.** Students read a passage related to the section topic and answer questions. **(GENERAL)**

_ **Directed Reading Worksheet, Characteristics of Fungi, CRF.** Students complete the exercises in this worksheet to help them understand the material as they read the section. **(BASIC)**

Block 3

TEACH *(30 minutes)*

_ **Integrating Physics and Chemistry**, TE. Students determine if fungi depends on physical changes, chemical changes, or both kinds of changes when obtaining nutrients. **(GENERAL)**

_ **Teaching Transparency, Penicillium Mold.** Use this transparency to discuss the structure of this common fungi. **(GENERAL)**

_ **Quick Lab,** Observing Bread Mold, SE. Students use a microscope to see the individual threads of cells that make up the body of a fungus. **(GENERAL)**

_ **Datasheets for In-Text Labs, Observing Bread Mold, CRF.**

CLOSE *(15 minutes)*

_ **Reteaching,** TE. Students compete in teams to review the section's objectives. **(BASIC)**

_ **Quiz,** TE. Students answer questions that review the section material. **(GENERAL)**

HOMEWORK

_ **Section Review,** SE. Assign questions 1–7 for review, homework, or quiz. **(GENERAL)**

_ **Quiz,** CRF. This quiz consists of ten multiple choice and matching questions that review the section's main concepts. **(BASIC) Also in Spanish.**

Other Resource Options

_ **Internet Connect.** Students can research Internet sources about Characteristics of Mushrooms with SciLinks Code HX4037.

_ **go.hrw.com.** For worksheets, videos, and other teaching aids related to this chapter, visit the HRW Web site and type in the keyword HX4 FNG.

_ **CNN Science in the News, Video Segment 18 Lethal Mushrooms.** This video segment is accompanied by a **Critical Thinking Worksheet**.

_ **CNN Student News.** Find the latest news, lesson plans, and activities related to important scientific events at **cnnstudentnews.com**.

Lesson Plan

Section: Fungal Diversity

Pacing

Regular Schedule:	**with lab(s):** 4 days	**without lab(s):** 2 days
Block Schedule:	**with lab(s):** 2 days	**without lab(s):** 1 day

Objectives

1. Describe the characteristics used to classify fungi.
2. List two commercial uses for fungi.
3. Describe three phyla of fungi.
4. Distinguish between the life cycles of zygomycetes, ascomycetes, and basidiomycetes.
5. Describe the mushroom *Amanita muscaria*.

National Science Education Standards Covered

UNIFYING CONCEPTS AND PROCESSES

UCP5: Form and function

SCIENCE AS INQUIRY

SI1: Abilities necessary to do scientific inquiry

SI2: Understandings about scientific inquiry

HISTORY AND NATURE OF SCIENCE

HNS1: Science as a human endeavor

LIFE SCIENCE: THE CELL

LSCell1: Cells have particular structures that underlie their functions.

LSCell2: Most cell functions involve chemical reactions.

LIFE SCIENCE: THE MOLECULAR BASIS OF HEREDITY

LSGene1: In all organisms, the instructions for specifying the characteristics of the organisms are carried in DNA.

LIFE SCIENCE: BIOLOGICAL EVOLUTION

LSEvol5: Biological classifications are based on how organisms are related.

LIFE SCIENCE: INTERDEPENDENCE OF ORGANISMS

LSInter3: Organisms both cooperate and compete in ecosystems.

LSInter5: Human beings live within the world's ecosystems.

KEY	
SE = Student Edition	**TE** = Teacher Edition
CRF = Chapter Resource File	

Block 4

FOCUS *(5 minutes)*

_ **Bellringer Transparency.** Use this transparency as students enter the classroom and find their seats. **(GENERAL)**

MOTIVATE *(10 minutes)*

_ **Discussion**, TE. Students list and describe foods made with the help of fungi.

TEACH *(30 minutes)*

_ **Teaching Transparency, Section Outline.** Use this transparency to give students a framework for the information in this section. **(GENERAL)**

_ **Teaching Transparency, Life Cycle of Zygomycetes.** Use this transparency to discuss the life cycle of *Rhizopus*, a black bread mold. Review the terms *hypha, zygosporangium, sporangium,* and *spore.* Highlight the points at which fusion, meiosis, and spore release take place. **(GENERAL)**

_ **Teaching Tip,** Hyphae, TE. Students make a drawing of *Rhizopus* and label the stolons and rhizoids. **(GENERAL)**

_ **Using the Figure,** Figure 6, TE. Students trace the life cycle of the mold shown in the figure. **(GENERAL)**

_ **Teaching Transparency, Life Cycle of Ascomycetes.** Use this transparency to discuss the life cycle of a typical ascomycete. Review the terms *ascus* and *ascocarp.* Point out that ascomycetes have septate hyphae. Tell students that ascomycetes are a diverse group that includes yeast, mildews, and cup fungi. **(GENERAL)**

HOMEWORK

_ **Directed Reading Worksheet, Fungal Diversity, CRF.** Students complete the exercises in this worksheet to help them understand the material as they read the section. **(BASIC)**

_ **Active Reading Worksheet, Fungal Diversity, CRF.** Students read a passage related to the section topic and answer questions. **(GENERAL)**

_ **Teaching Tip,** Comparing Fungi, TE. Students compare the three sexually-reproducing phyla of fungi. **(ADVANCED)**

Block 5

TEACH *(30 minutes)*

_ **Teaching Transparency, Life Cycle of Basdiomycetes.** Use this transparency to discuss the life cycle of a typical basidiomycete. Point out that the main part of the fungi is below ground. **(GENERAL)**

_ **Teaching Transparency, Up Close Mushroom.** Use this transparency to discuss the mushroom *Amanita muscaria*, its main structures, and their functions. **(GENERAL)**

_ **Up Close**, Mushroom, TE. Use these teaching strategies and discussion questions to help students better understand the information provided along with the art in the SE.

_ **Inclusion Strategies**, TE. Students explore the use of mushrooms in dishes through restaurant menus, by interviewing cooks, and through their own family's recipes.

CLOSE *(15 minutes)*

_ **Reteaching,** TE. Students collect mushrooms and identify them using field guides. **(BASIC)**

_ **Quiz,** TE. Students answer questions that review the section material. **(GENERAL)**

HOMEWORK

_ **Alternative Assessment**, TE. Students make flashcards of fungal structures and their associated phyla. **(GENERAL)**

_ **Quiz, CRF.** This quiz consists of ten multiple choice and matching questions that review the section's main concepts. **(BASIC) Also in Spanish.**

_ **Section Review,** SE. Assign questions 1–5 for review, homework, or quiz. **(GENERAL)**

Optional Blocks

LABS *(90 minutes)*

_ **Exploration Lab, Exploring Fungi, CRF.** Students examine and compare specimens from the four fungi phyla. **(GENERAL)**

_ **Skills Practice Lab, Best Food for Yeast, CRF.** Students use different sugars to test how well one kind of yeast carries out fermentation. They use a CO_2 gas sensor to measure the amount of carbon dioxide produced during fermentation. **(GENERAL)**

Other Resource Options

- **Internet Connect.** Students can research Internet sources about Characteristics of Mushrooms with SciLinks Code HX4037.
- **go.hrw.com.** For worksheets, videos, and other teaching aids related to this chapter, visit the HRW Web site and type in the keyword HX4 FNG.
- **CNN Science in the News, Video Segment 18 Lethal Mushrooms.** This video segment is accompanied by a **Critical Thinking Worksheet**.
- **CNN Student News.** Find the latest news, lesson plans, and activities related to important scientific events at **cnnstudentnews.com**.

Lesson Plan

Section: Fungal Partnerships

Pacing

Regular Schedule:	**with lab(s):** 3 days	**without lab(s):** 2 days
Block Schedule:	**with lab(s):** 1 1/2 days	**without lab(s):** 1 day

Objectives

1. Distinguish two symbiotic relationships that involve fungi.
2. Summarize the ecological importance of mycorrhizae.
3. Describe lichens.

National Science Education Standards Covered

UCP5: Form and function

SCIENCE AS INQUIRY

SI1: Abilities necessary to do scientific inquiry

SI2: Understandings about scientific inquiry

HISTORY AND NATURE OF SCIENCE

HNS1: Science as a human endeavor

LIFE SCIENCE: THE CELL

LSCell1: Cells have particular structures that underlie their functions.

LSCell2: Most cell functions involve chemical reactions.

LIFE SCIENCE: THE MOLECULAR BASIS OF HEREDITY

LSGene1: In all organisms, the instructions for specifying the characteristics of the organisms are carried in DNA.

LIFE SCIENCE: BIOLOGICAL EVOLUTION

LSEvol5: Biological classifications are based on how organisms are related.

LIFE SCIENCE: INTERDEPENDENCE OF ORGANISMS

LSInter3: Organisms both cooperate and compete in ecosystems.

KEY

SE = Student Edition **TE** = Teacher Edition

CRF = Chapter Resource File

Block 6

FOCUS *(5 minutes)*

_ **Bellringer Transparency.** Use this transparency as students enter the classroom and find their seats. **(GENERAL)**

MOTIVATE *(10 minutes)*

_ **Demonstration**, TE. Students describe pictures or specimens of different common forms of lichen. **(BASIC)**

TEACH *(30 minutes)*

_ **Teaching Transparency, Section Outline.** Use this transparency to give students a framework for the information in this section. **(GENERAL)**

_ **Directed Reading Worksheet, Fungal Partnerships, CRF.** Students complete the exercises in this worksheet to help them understand the material as they read the section. **(BASIC)**

_ **Using the Figure**, Figure 10, TE. Use the figure to explain how mycorrhizae increase the absorbing capacity of plant roots. **(GENERAL)**

HOMEWORK

_ **Active Reading Worksheet, Fungal Partnerships, CRF.** Students read a passage related to the section topic and answer questions. **(GENERAL)**

Block 7

TEACH *(25 minutes)*

_ **Data Lab,** Analyzing the Effect of Mycorrhizae, SE. Students study lab data on plant growth to discover the cause of differences in the growth rate. **(GENERAL)**

_ **Datasheets for In-Text Labs, Analyzing the Effect of Mycorrhizae, CRF.**

_ **Exploring Further,** Lichens as Environmental Watchdogs, SE. Students read the article and then do research to find out more about indicator species. **(GENERAL)**

CLOSE *(20 minutes)*

_ **Science Skills Worksheet, CRF.** Students complete a table that compares an oak tree with a mushroom. **(GENERAL)**

- **Reteaching,** TE. Students construct a two-column table to compare lichens and mycorrhizae. **(BASIC)**
- **Quiz,** TE. Students answer questions that review the section material. **(GENERAL)**

HOMEWORK

- **Section Review,** SE. Assign questions 1–5 for review, homework, or quiz. **(GENERAL)**
- **Quiz, CRF.** This quiz consists of ten multiple choice and matching questions that review the section's main concepts. **(BASIC) Also in Spanish.**
- **Modified Worksheet, One-Stop Planner.** This worksheet has been specially modified to reach struggling students. **(BASIC)**
- **Critical Thinking Worksheet, CRF.** Students answer analogy-based questions that review the section's main concepts and vocabulary. **(ADVANCED)**

Optional Block

LAB *(45 minutes)*

- **Skills Practice Lab,** Observing Yeast and Fermentation, SE. Students observe and measure the products of fermentation. **(GENERAL)**
- **Datasheets for In-Text Labs, Observing Yeast and Fermentation, CRF.**

Other Resource Options

- **Supplemental Reading, The Andromeda Strain, One-Stop Planner.** Students read the book and answer questions. **(ADVANCED)**
- **Career,** Plant Pathologist, TE. Discuss the job and importance of a plant pathologist.
- **Internet Connect.** Students can research Internet sources about Symbiosis of Fungi with SciLinks Code HX4172.
- **Internet Connect.** Students can research Internet sources about Lichens with SciLinks Code HX4114.
- **go.hrw.com.** For worksheets, videos, and other teaching aids related to this chapter, visit the HRW Web site and type in the keyword HX4 FNG.
- **CNN Science in the News, Video Segment 18 Lethal Mushrooms.** This video segment is accompanied by a **Critical Thinking Worksheet**.
- **CNN Student News.** Find the latest news, lesson plans, and activities related to important scientific events at **cnnstudentnews.com**.

Lesson Plan

End-of-Chapter Review and Assessment

Pacing

Regular Schedule: 2 days

Block Schedule: 1 day

KEY	
SE = Student Edition	**TE** = Teacher Edition
CRF = Chapter Resource File	

Block 8

REVIEW *(45 minutes)*

- **Study Zone,** SE. Use the Study Zone to review the Key Concepts and Key Terms of the chapter and prepare students for the Performance Zone questions. **(GENERAL)**
- **Performance Zone,** SE. Assign questions to review the material for this chapter. Use the assignment guide to customize review for sections covered. **(GENERAL)**
- **Teaching Transparency, Concept Mapping.** Use this transparency to review the concept map for this chapter. **(GENERAL)**

Block 9

ASSESSMENT *(45 minutes)*

- **Chapter Test, Fungi, CRF.** This test contains 20 multiple choice and matching questions keyed to the chapter's objectives. **(GENERAL) Also in Spanish.**
- **Chapter Test, Fungi, CRF.** This test contains 25 questions of various formats, each keyed to the chapter's objectives. **(ADVANCED)**
- **Modified Chapter Test, One-Stop Planner.** This test has been specially modified to reach struggling students. **(BASIC)**

Other Resource Options

- **Vocabulary Review Worksheet, CRF.** Use this worksheet to review the chapter vocabulary. **(GENERAL) Also in Spanish.**
- **Test Prep Pretest, CRF.** Use this pretest to review the main content of the chapter. Each question is keyed to a section objective. **(GENERAL) Also in Spanish.**
- **Test Item Listing for ExamView® Test Generator, CRF.** Use the Test Item Listing to identify questions to use in a customized homework, quiz, or test.
- **ExamView® Test Generator, One-Stop Planner.** Create a customized homework, quiz, or test using the HRW Test Generator program.

TEST ITEM LISTING

Fungi

TRUE/FALSE

1. ____ All fungi are heterotrophic.
Answer: True Difficulty: I Section: 1 Objective: 1

2. ____ The cell walls of fungi are made up of cellulose.
Answer: False Difficulty: I Section: 1 Objective: 1

3. ____ All fungi except yeasts have bodies composed of filaments.
Answer: True Difficulty: I Section: 1 Objective: 2

4. ____ Some fungi are parasites.
Answer: True Difficulty: I Section: 1 Objective: 3

5. ____ Fungi obtain nutrients through photosynthesis.
Answer: False Difficulty: I Section: 1 Objective: 3

6. ____ Hyphae are slender filaments that increase the surface area-to-volume ratio of a fungus, enabling the fungus to absorb nutrients efficiently from the environment.
Answer: True Difficulty: I Section: 1 Objective: 4

7. ____ Fungal spores are haploid spores formed by meiosis during sexual reproduction.
Answer: False Difficulty: I Section: 1 Objective: 5

8. ____ The classification of organisms in the three phyla of the kingdom Fungi is based on sexual reproductive structure.
Answer: True Difficulty: I Section: 2 Objective: 1

9. ____ Commercial uses for fungi include antibiotics, wine-making, and cheese-making.
Answer: True Difficulty: I Section: 2 Objective: 2

10. ____ Some yeasts can cause diseases in humans.
Answer: True Difficulty: I Section: 2 Objective: 3

11. ____ Sporangia are reproductive structures in which spores form.
Answer: True Difficulty: I Section: 2 Objective: 4

12. ____ The recognizable part of *Amanita muscaria* is the mushroom.
Answer: True Difficulty: I Section: 2 Objective: 5

13. ____ The basidium is a club-shaped nutrient-collecting structure.
Answer: False Difficulty: I Section: 2 Objective: 5

14. ____ Fungi usually form symbiotic associations with animals and bacteria.
Answer: False Difficulty: I Section: 3 Objective: 1

15. ____ Some fungi aid in the transfer of minerals from the soil to the roots of plants.
Answer: True Difficulty: I Section: 3 Objective: 2

16. ____ Lichens have been found all over the world except in arid desert regions.
Answer: False Difficulty: I Section: 3 Objective: 3

MULTIPLE CHOICE

17. Fungi obtain energy
    a. directly from the sun.
    b. from inorganic material in their environment.
    c. by absorbing organic molecules.
    d. from nuclear fusion.

    Answer: C Difficulty: I Section: 1 Objective: 1

18. Chitin is found in fungi and in
    a. clam shells.
    b. the outer shells of insects.
    c. some plant cell walls.
    d. snail shells.

    Answer: B Difficulty: I Section: 1 Objective: 1

19. Fungi
    a. do not contain chlorophyll.
    b. have cell walls that contain chitin.
    c. do not produce their own food.
    d. All of the above

    Answer: D Difficulty: I Section: 1 Objective: 1

20. A hypha is a long string of cells divided by
    a. spindle fibers
    b. an ascus
    c. mycorrhizae
    d. walls

    Answer: D Difficulty: I Section: 1 Objective: 2

21. The individual filaments that make up the body of a fungus are called
    a. vascular tissue.
    b. hyphae.
    c. rhizoids.
    d. stem cells.

    Answer: B Difficulty: I Section: 1 Objective: 2

22. fungal food : organic molecules ::
    a. pea plant cell wall : chitin
    b. insect exoskeleton : cellulose
    c. fungus cell wall : cellulose
    d. mycelium : hyphae

    Answer: D Difficulty: II Section: 1 Objective: 2

23. Fungi obtain food by
    a. photosynthesis.
    b. the nitrogen fixation process in their hyphae.
    c. digesting food externally before absorbing it.
    d. None of the above

    Answer: C Difficulty: I Section: 1 Objective: 3

24. Fungi digest food
    a. through photosynthesis.
    b. outside their bodies.
    c. inside their bodies.
    d. All of the above

    Answer: B Difficulty: I Section: 1 Objective: 3

25. Fungi are important to an ecosystem as
    a. producers.
    b. regulators.
    c. decomposers.
    d. controllers.

    Answer: C Difficulty: I Section: 1 Objective: 4

26. Most fungal spores are formed by
    a. the fusing of hyphae.
    b. the fusing of asci.
    c. mitosis.
    d. None of the above

    Answer: C Difficulty: I Section: 1 Objective: 5

27. An economically important use of fungi is
    a. bread-making.
    b. the production of antibiotics.
    c. the manufacture of alcoholic beverages.
    d. All of the above

    Answer: D   Difficulty: I   Section: 2   Objective: 2

28. yeasts : bread ::
    a. hyphae : wine-making
    b. aspergillus : soy sauce
    c. molds : drug manufacturing
    d. ascus : making cheeses

    Answer: B   Difficulty: II   Section: 2   Objective: 2

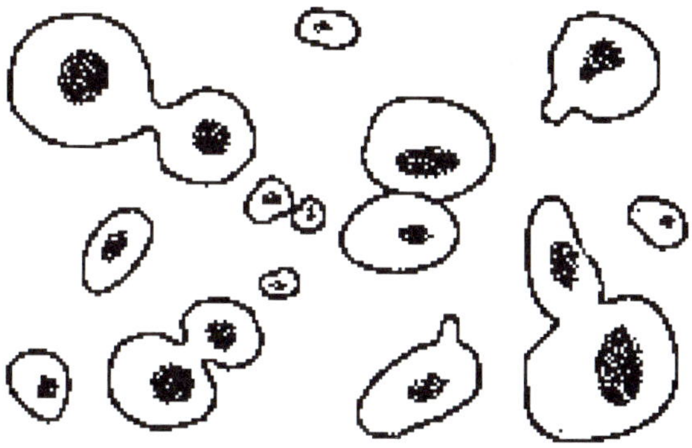

29. Refer to the illustration above. The organism is a(n)
    a. mycorrhizae.
    b. zygomycete.
    c. basidiomycete.
    d. unicellular fungus.

    Answer: D   Difficulty: II   Section: 2   Objective: 3

30. Refer to the illustration above. The organism can reproduce
    a. by fission.
    b. by forming buds.
    c. sexually.
    d. All of the above

    Answer: D   Difficulty: II   Section: 2   Objective: 3

31. Mushrooms, puffballs, and shelf fungi are examples of
    a. club fungi.
    b. sac fungi.
    c. molds.
    d. yeasts.

    Answer: A   Difficulty: I   Section: 2   Objective: 3

32. Mushrooms and toadstools are members of the phylum
    a. Ascomycota.
    b. Basidiomycota.
    c. Zygomycota.
    d. Deuteromycota.

    Answer: B   Difficulty: I   Section: 2   Objective: 3

33. An example of a fungus is
    a. a mushroom.
    b. a bread mold.
    c. a yeast.
    d. All of the above

    Answer: D   Difficulty: I   Section: 2   Objective: 3

34. All of the following are true of ascomycetes *except*
    a. they have saclike reproductive cells in which spores grow.
    b. some are truffles.
    c. they are often used in the baking of bread.
    d. no sexual stage has been seen.

    Answer: D   Difficulty: I   Section: 2   Objective: 3

35. The group of fungi that includes the molds that often grow on bread is the
    a. ascomycetes.
    b. basidiomycetes.
    c. zygomycetes.
    d. deuteromycetes.

    Answer: C   Difficulty: I   Section: 2   Objective: 3

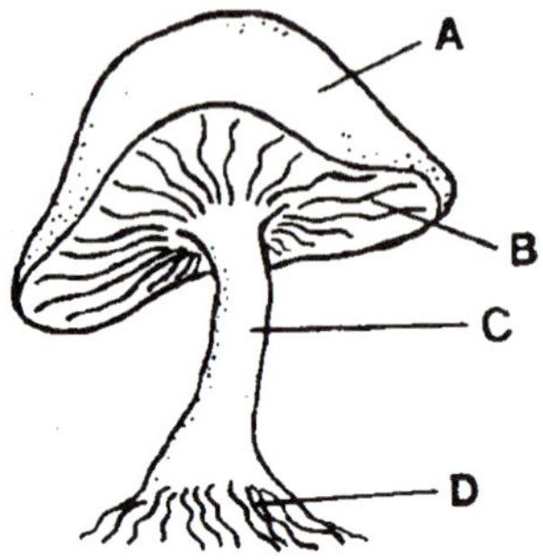

36. Refer to the illustration above. Which features characterize the kingdom of which this organism is a member?
    a. eukaryotic, absorbs nutrients
    b. aquatic, multicellular
    c. autotrophic, ingests nutrients
    d. prokaryotic, photosynthetic

    Answer: A   Difficulty: II   Section: 2   Objective: 3

37. Refer to the illustration above. Which structure is responsible for meeting the food requirements of the organism shown?
    a. A
    b. B
    c. C
    d. D

    Answer: D   Difficulty: II   Section: 2   Objective: 4

38. Zygospores allow molds to
    a. remain dormant until conditions are favorable for their spores.
    b. digest bread.
    c. grow unusually large.
    d. produce antibiotics.

    Answer: A   Difficulty: I   Section: 2   Objective: 4

39. Reproductive structures in which spores form are known as
    a. septa.
    b. cnidophores.
    c. mycorrhizae.
    d. sporangia.

    Answer: D   Difficulty: I   Section: 2   Objective: 4

Life Cycle of a Mold

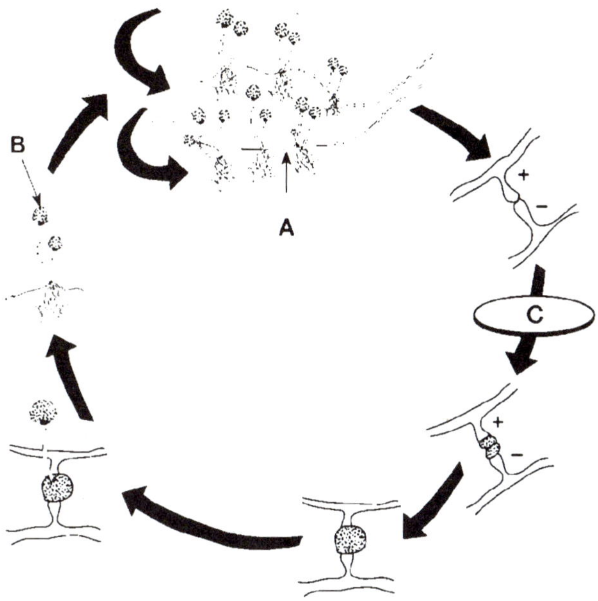

40. Refer to the illustration above. Structure A is
    a. a rhizoid.
    b. vascular tissue.
    c. a hypha.
    d. a stem.

    Answer: C Difficulty: II Section: 2 Objective: 4

41. Refer to the illustration above. Structure B is
    a. commensal.
    b. haploid.
    c. embryonic.
    d. diploid.

    Answer: B Difficulty: II Section: 2 Objective: 4

42. Refer to the illustration above. The process that takes place at C is known as
    a. meiosis.
    b. conjugation.
    c. mitosis.
    d. fusion.

    Answer: D Difficulty: II Section: 2 Objective: 4

43. In a symbiotic association, such as a lichen, a fungus provides mineral nutrients to a(n)
    a. animal.
    b. heterotrophic bacterium.
    c. photosynthetic partner.
    d. None of the above

    Answer: C Difficulty: I Section: 3 Objective: 1

44. Mycorrhizae
    a. aid in the transfer of minerals from the soil to a plant.
    b. cause a variety of plant diseases.
    c. aid in the transfer of minerals to fungi.
    d. are only found on aquatic fungi.

    Answer: A Difficulty: I Section: 3 Objective: 2

45. Some mycorrhizae do *not* penetrate the host plant but rather wrap around the
    a. stem.
    b. cap.
    c. root.
    d. gills.

    Answer: C Difficulty: I Section: 3 Objective: 2

46. A lichen
    a. consists of a fungus and an alga in a symbiotic relationship.
    b. is a fungus clump.
    c. is found only in temperate climates.
    d. is a mold found on the shady side of trees.

    Answer: A   Difficulty: I   Section: 3   Objective: 3

47. fungus : lichen ::
    a. plant : mycorrhiza
    b. mycelium : hypha
    c. walls : hypha
    d. mycelium : fungus

    Answer: A   Difficulty: II   Section: 3   Objective: 3

COMPLETION

48. Unlike plants, fungi lack ____________________ and cannot carry out photosynthesis.
    Answer: chloroplasts   Difficulty: I   Section: 1   Objective: 1

49. The typical fungus is a eukaryotic heterotroph that has a body consisting of many slender filaments called ____________________.
    Answer: hyphae   Difficulty: I   Section: 1   Objective: 2

50. When hyphae grow, they form a mass called a(n) ____________________.
    Answer: mycelium   Difficulty: I   Section: 1   Objective: 2

51. Fungi obtain food by ____________________ organic matter.
    Answer: decomposing   Difficulty: I   Section: 1   Objective: 3

52. Fungi that absorb nutrients in a person's body can cause life-threatening ____________________.
    Answer: infections   Difficulty: I   Section: 1   Objective: 4

53. Fungi reproduce sexually and ____________________.
    Answer: asexually   Difficulty: I   Section: 1   Objective: 5

54. Most fungal spores are formed by ____________________.
    Answer: mitosis   Difficulty: I   Section: 2   Objective: 1

55. Chestnut blight is caused by fungi belonging to the phylum called ____________________.
    Answer: Ascomycota   Difficulty: II   Section: 2   Objective: 3

56. The familiar mushroom belongs to the phyla ____________________.
    Answer: Basidiomycota   Difficulty: II   Section: 2   Objective: 3

57. Mildews and yeasts are examples of ____________________.
    Answer: ascomycetes   Difficulty: I   Section: 2   Objective: 3

58. The common name given to unicellular ascomycetes is ____________________.
    Answer: yeasts   Difficulty: I   Section: 2   Objective: 4

59. A(n) ____________________ is a saclike structure in which haploid spores are formed.
    Answer: ascus   Difficulty: I   Section: 2   Objective: 4

60. A fungal ____________________ is a haploid reproductive cell that is capable of developing into a new organism.
    Answer: spore   Difficulty: I   Section: 2   Objective: 4

61. Fungi form symbiotic relationships with a(n) ____________________, such as a plant or an alga.
    Answer: photosynthesizer   Difficulty: I   Section: 3   Objective: 1

62. Certain fungi play important roles in the nutrition of vascular plants by forming symbiotic associations with their roots, which are called ____________________.
Answer: mycorrhizae Difficulty: I Section: 3 Objective: 2

63. Lichens are able to carry out ____________________ fixation.
Answer: nitrogen Difficulty: I Section: 3 Objective: 3

64. Lichens are sensitive to ____________________ such as sulfur dioxide in the atmosphere.
Answer: pollutants Difficulty: I Section: 3 Objective: 3

65. A lichen consists of a fungus and a(n) ____________________ living together in a symbiotic relationship.
Answer: alga Difficulty: I Section: 3 Objective: 3

ESSAY

66. How does mitosis in a mushroom differ from that in most other eukaryotic organisms?
Answer:
Fungi exhibit nuclear mitosis. In most organisms, the nuclear membrane breaks down in the early stages of mitosis. In dividing mushroom cells, however, the nuclear envelope remains intact from prophase to anaphase. The spindle fibers form within the nucleus and drag the chromosomes to opposite poles of the nucleus, rather than to opposite poles of the cell. Mitosis is complete when the nuclear envelope pinches in two. This basic difference in the process of cell division suggests that fungi are not closely related to any other organism.
Difficulty: III Section: 1 Objective: 1

67. Why are fungi well suited for absorbing food from the environment?
Answer:
The mycelium of a fungus can be made of many meters of individual hyphae. This creates a high surface area-to-volume ratio, which makes a fungus well suited for absorbing food from the environment.
Difficulty: II Section: 1 Objective: 3

68. Explain how, under different conditions, either sexual or asexual reproduction might be advantageous to a fungus.
Answer:
During periods when the environment is favorable, the rapid asexual formation of spores ensures the quick spread of the species. In periods of environmental stress, sexual reproduction ensures genetic recombination before the hyphae die.
Difficulty: II Section: 2 Objective: 4

69. Why are lichens a good indicator of the environment's health?
Answer:
Lichens are particularly sensitive to pollutants in the atmosphere because they readily absorb substances dissolved in rain and dew. Lichen are some of the first living organisms to be destroyed by air pollutants and can warn us to curb pollutants before other species are destroyed.
Difficulty: II Section: 3 Objective: 3

70. Describe the symbiotic relationship that enables a lichen to exist.
Answer:
A lichen is a symbiosis between a fungus and a photosynthetic partner such as a green alga, a cyanobacterium, or both. The fungus absorbs minerals and other nutrients from rocks or logs and retains water that the photosynthetic partner needs for photosynthesis. Through photosynthesis, the alga produces carbohydrates that the fungus uses for food.
Difficulty: III Section: 3 Objective: 3

9997266005 1 2 3 4 5 6